Rotenone Use in Fisheries Management:

Administrative and Technical Guidelines Manual

This document was made possible by funds provided by the

U.S. Fish and Wildlife Service,
Division of Federal Aid

Rotenone Use in Fisheries Management

American Fisheries Society
Fish Management Chemicals Subcommittee
Task Force on Fishery Chemicals

Authors

Brian J. Finlayson*, California Department of Fish and Game
Rosalie A. Schnick, Michigan State University
Richard L. Cailteux, Florida Fish and Wildlife Conservation Commission
Leo DeMong, New York Department of Environmental Conservation
William D. Horton, Idaho Department of Fish and Game
William McClay, Michigan Department of Natural Resources
Charles W. Thompson, Utah Division of Wildlife Resources
Gregory J. Tichacek, Illinois Department of Natural Resources

*Author to which correspondence should be submitted:
Brian Finlayson
California Department of Fish and Game
1701 Nimbus Road, Suite F
Rancho Cordova, CA 95670

Suggested citation formats follow.

Entire book

Finlayson, B. J., R. A. Schnick, R. L. Cailteux, L. DeMong, W. D. Horton, W. McClay, C. W. Thompson, and G. J. Tichacek. 2000. Rotenone use in fisheries management: administrative and technical guidelines manual. American Fisheries Society, Bethesda, Maryland.

Library of Congress Catalog Card Number 00-101915
ISBN 1-888569-22-0

American Fisheries Society
5410 Grosvenor Lane, Suite 110
Bethesda, Maryland 20814-2199
USA

Contents

ACKNOWLEDGMENTS

This document was made possible by funds provided by the U.S. Fish and Wildlife Service, Division of Federal Aid (Administrative Grants AP98-012). The authors express their gratitude to Paul Brouha (past AFS Executive Director) for his endless and undaunted enthusiasm and support obtaining the grant. The authors would like to express their gratitude for comments and suggestions received from the following reviewers: Don Archer (Utah Division of Wildlife Resources), Brian Chan (British Columbia Ministry of Environment, Lands and Parks), Joe Conti and Ruth Fisher (AgrEvo Environmental Health), Richard Flint (California Department of Fish and Game), Sherman Hebein (Colorado Division of Wildlife), David Johnson (U.S. Fish and Wildlife Service), Larry Nashett (New York State Department of Environmental Conservation), Hal Schramm (U.S. Geological Survey, Mississippi Cooperative Fish and Wildlife Research Unit), Gerry Taylor (retired), William Taylor and Edward Roseman (Michigan State University), Ray Temple (U.S. Fish and Wildlife Service), and Jerry Weigel (North Dakota Game and Fish Department). The section on applicator safety was written by Joel Trumbo (California Department of Fish and Game).

Executive Summary

Rotenone Use in Fisheries Management: Administrative and Technical Guidelines Manual was made possible by funds provided in December 1997 by the U.S. Fish and Wildlife Service, Division of Federal Aid (Administrative Grants AP98-012), and requested by the American Fisheries Society's Task Force on Fishery Chemicals in a proposal entitled the "Rotenone Stewardship Program." This manual was written to detail the proper use of rotenone for fish biologists and fishery managers in the United States and Canada and has been distributed to fish and wildlife agencies in the United States and Canada. More specifically, the objectives of this manual are to (1) promote the continued safe and effective use of rotenone as a piscicide and as a sampling tool of fish populations and (2) make specific recommendations for the administrative and technical procedures for rotenone applications that will ensure continued availability of this valuable fish management tool. Information is provided on planning and executing a treatment project and for addressing issues identified by fish biologists in a survey conducted by the authors of this manual.

Despite the importance of rotenone in fisheries management, its continued availability and use are uncertain. The majority of rotenone treatments have occurred without incident. However, a small number of treatments have generated widespread public controversy. As more demands are placed on water bodies and the public becomes more environmentally aware, fisheries professionals must respond with guidelines for using rotenone prudently with minimal impacts. The authors of this manual believe that most of the incidents with rotenone could have been avoided if the responsible agency had (1) garnered more public input and support prior to the treatment and not been in an adversarial role with the local community, (2) done a better job of implementing the treatment with appropriate procedures and qualified personnel, and (3) provided better technical, administrative, legal, and political support after the treatment. The manual focuses on correcting these three deficiencies, recognizing that, if left uncorrected, these will eventually result in the loss of this critical management tool.

The manual is divided into six sections that provide information on the following topics:

Section 1, Introduction, contains (a) background information on rotenone, (b) the history of rotenone use in fisheries management in the United States and Canada, (c) rotenone use in controlling and eradicating populations of fish, (d) advantages and limitations of rotenone and other control methods, (e) rotenone use in sampling fish communities, (f) advantages and limitations of rotenone and other sampling methods, (g) current legal and environmental constraints on rotenone use in fisheries management, (h) status of registration, and (i) an explanation of terminology and glossary of terms.

Section 2, Administrative Procedures, contains a stepwise scheme for (a) preliminary planning, where the project concept is developed and public input and acceptance are initiated to assess the resources needed, (b) intermediate planning, where an environmental analysis refines the project and pubic acceptance is gained, (c) implementation and management, where project-specific work plans are developed, and (d) performing the treatment. Other important issues discussed include the value of using fisheries management plans for water bodies and species, techniques for gaining public input and acceptance of rotenone projects, and techniques for managing crises.

Section 3, Technical Procedures, contains information on (a) techniques and equipment (including application charts) for treatment of ponds, lakes, and reservoirs, (b) techniques and equipment (including application charts) for treatment of streams and rivers, (c) safety for applicators using liquid and powder formulations, (d) monitoring procedures for air, water, and sediment, (e) neutralization procedures for rotenone, and (f) fish collection and disposal. This section also contains the latest rotenone labels and monographs and numerous photographs and diagrams of application equipment and techniques.

Section 4, Project Assessment, contains recommendations for assessing the short- and long- term success and impact of a rotenone treatment.

Section 5, Issues and Responses, contains answers to common popular questions about rotenone use. This stand-alone section contains general information and data on the impact of rotenone on public health, environmental quality, and fish and wildlife. References are provided at the end of this section (as well as in Section 6) so the entire section can be independently duplicated and distributed to the public. This section will be updated annually on the AFS web site.

Section 6, References, contains full citations on the literature cited in the text.

Mention of trade names or commercial products does not constitute endorsement or recommendation for use. The manual was intended to provide fishery professionals with guidelines for rotenone use, and these views do not necessarily reflect those of the American Fisheries Society, the U.S. Fish and Wildlife Service, or the individual employers of the authors.

The Rotenone Stewardship Program also includes a public information program to educate the public on the benefits and risks of rotenone use and an electronic information system for fisheries biologists that will provide up-to-date information on current use restrictions, experts on the use of rotenone, important issues and solutions, and the registration status of rotenone.

Common Abbreviations

AF	acre-feet
AFS	American Fisheries Society
CDFG	California Department of Fish and Game
EA	environmental assessment, environmental analysis
EIS	environmental impact statement
FEPCA	Federal Environmental Pesticide Control Act
FMP	fisheries management plan
FONSI	finding of no significant impact
GPS	Global Positioning System
GCMS	gas chromatograph/mass spectroscopy
HPLC	high performance liquid chromatography
ICS	incident command system
$KMnO_4$	potassium permanganate
NEPA	National Environmental Policy Act
NOEL	no observable effect level
OSHA	Occupational Safety and Health Administration
PBO	piperonyl butoxide
PCPA	Pest Control Products Act
PEL	permissible exposure limits
PIP	public involvement plan
PMRA	Pest Management Regulatory Agency (Canada)
ppb	parts per billion (equivalent to µg/L or µg/kg)
ppm	parts per million (equivalent to mg/L or mg/kg)
semiVOC	semivolatile organic compounds
SNARL	suggested no-adverse response level
TCE	trichloroethylene
UDWR	Utah Division of Wildlife Resources
UMESC	Upper Midwest Environmental Sciences Center
USEPA	U.S. Environmental Protection Agency
USFWS	U.S. Fish and Wildlife Service
USGS	U.S. Geological Survey
VOC	volatile organic compounds

Introduction 1

Fisheries managers rely on a wide variety of tools for the management and assessment of fish populations to maintain diverse and productive aquatic ecosystems and quality recreational fisheries. One of the most valuable tools is the piscicide, rotenone. Piscicide application is the only method that can completely eradicate undesirable fish communities or accurately sample a portion of the entire fish population, including all species. Despite the ongoing need for rotenone, its continued use remains uncertain. Over the past several years, the use of rotenone has become a concern to environmental and animal rights groups. As a result, its use has been challenged, halted, or discouraged (B. Finlayson and R. A. Schnick, American Fisheries Society proposal for the use of Federal Aid Administration funds to the U.S. Fish and Wildlife Service, unpublished, 1996).

The objectives of this manual are to (1) promote the continued safe and effective use of rotenone as a piscicide for sampling or controlling fish populations; and (2) make specific recommendations for the administrative and technical procedures for rotenone applications to ensure the continued availability of this valuable fish management tool.

Rotenone is found in Australia, Oceania, southern Asia, and South America as a naturally occurring substance derived from the roots of tropical plants in the bean family (Leguminosae) including jewel vine (*Derris* spp.) and lacepod (*Lonchocarpus* spp.). Rotenone has been used for centuries to capture fish in areas where these plants are naturally found. Fisheries managers in North America began to use rotenone for fisheries management in the 1930s. The piscicide was applied first to ponds and lakes, and then to streams in the early 1960s for either complete or partial reclamation (Schnick 1974). Rotenone was initially used in various powdered forms until wettable pastes and emulsifiable formulations were developed that acted faster and were easier to handle and dispense. By 1949, 34 states and several Canadian provinces were using rotenone routinely for the management of fish populations (Solman

1950; Lennon et al. 1970). Rotenone is also used to sample fish populations for management purposes and to act as a natural insecticide for use on crops and livestock. Humans have taken rotenone orally to control intestinal worms (Haley 1978).

In 1993, the American Fisheries Society (AFS) recognized a need to respond to increased concerns about rotenone use from environmental and animal rights groups. Response information was needed to supplement a U.S. Fish and Wildlife Service (USFWS) brochure developed by Sousa et al. (1987a). This recognition led to a proposal for a "Rotenone Stewardship Program" from the AFS Task Force on Fishery Chemicals to the USFWS Division of Federal Aid to use administrative funds for the preparation and production of a manual on rotenone use by fisheries managers (Finlayson and Schnick, unpublished). The proposal was selected for two years of funding by USFWS on 2 December 1997.

1.1 CONTROL OR ERADICATION OF FISH POPULATIONS BY PISCICIDES

As many as 30 piscicides have been used extensively in fisheries management in the United States and Canada since the 1930s, but only four are currently registered for general or selective fish control or sampling. These products include the general piscicides, antimycin and rotenone, and the lampricides, Lamprecid® and Bayluscide®. Rotenone is the most extensively used piscicide in the United States (Cumming 1975).

1.1.1 Reasons for controlling or eradicating fish populations

Fisheries managers may decide to use rotenone when fish communities have been disrupted by human activities (e.g., physical manipulations of natural waters, effects of pollution on natural production of fish species, demand for recreational fisheries, and introduction of exotic species into surface waters). Only piscicide applications or complete dewatering can eradicate entire populations of undesirable fish (Schnick 1974).

The careful and professional use of piscicides in the United States was stimulated by the passage of the Dingell-Johnson Act in 1950 (Federal Aid in Sport Fish Restoration), establishment of the Cooperative Fishery Unit program in 1960, publication of training and manuals in the 1960s and 1970s, and development of sophisticated application equipment (Lennon et al. 1970).

The primary reasons for piscicide use have changed. Originally, piscicides were mainly used to control undesirable fish populations so that sport fish could be stocked and managed for recreational purposes in lakes, ponds, and streams without competition, predation, or other interference by the undesirable fish (Lennon et al. 1970; Finlayson and Schnick, unpublished). Today, the most frequently reported uses (in order of the amount of active ingredient used) are (1) control of undesirable fish to support recreational fisheries, (2) eradication of exotic fish, (3) eradication of competing fish species in rearing facilities or ponds, (4) quantification of populations of aquatic organisms, (5) treatment of drainages before initial reservoir impoundment, (6) eradication of fish to control disease, and (7) restoration of threatened or endangered species (see Appendix A).

1.1.2 Alternative control methods (Adapted from CDFG 1994)

Other methods for reducing or controlling fish communities include (1) use of a piscicide other than rotenone (antimycin registered as Fintrol® is the only other piscicide registered for general use in the United States), (2) modification of angling regulations (i.e., modifications to promote or favor overharvest), (3) physical removal techniques (e.g., nets, traps, or electrofishing), (4) biological control techniques (i.e., predators, intraspecific manipulation, pathological reactions), (5) dewatering or water fluc-

Table 1.1 Advantages and limitations of control methods.

Type of control	Advantages	Limitations
Piscicide—rotenone	•Except for dewatering, only control method for complete eradication of fish populations •Can be applied to achieve spatially selective eradications •Can be used in large river systems •Rapid results •Controls all post-embryonic life stages	•Temporary loss of potable water supplies and recreational opportunities •Temporary effects on aquatic habitat and nontarget species (e.g., zooplankton, newts) •Can be repellent •Does not kill fish eggs until the shell ruptures at hatching
Piscicide—antimycin	•Except for dewatering, only control method for complete eradication of fish populations •Controls all post-embryonic life stages •Can be selective by species •Nonrepellent •Rapid results	•Not registered in every state •Limited availability and history •Not effective at high pHs (≥8.5) •Does not kill fish eggs until the shell ruptures at hatching •Temporary loss of potable water supplies and recreational opportunities •Temporary effects on aquatic habitat and nontarget species (e.g., zooplankton)
Modifications of angling regulations	•Generally publicly acceptable •Low cost •Utilization as food	•Usually slow •Angler pressure inadequate •Often ineffective •Many species not vulnerable to angling
Physical removal techniques (nets, traps, or electrofishing)	•Publicly acceptable	•Need high exploitation rates •Juveniles and other game fish fill void •Expensive and labor intensive •Potential escapement •Benefits are of short duration
Biological control techniques (predators, intraspecific	•May be low cost	•Limited success in maintaining predator populations •Difficulties with techniques

Table 1.1 continued.

Type of control	Advantages	Limitations
manipulation, and pathological reactions)		•Unpredictable results •Inability to control introduced pathogens •Legal concerns
Dewatering and water fluctuation techniques	•Except for piscicides, only control method for a complete eradication of fish populations •May be low cost	•Water remains in same pools and stream sections •Can be detrimental to game fish •Environmentally disruptive
Streamflow augmentation techniques	•May be publicly acceptable •Easy to implement	•Need for controls to change flow •Water rights issues •Multiple-use conflicts
Fish barriers	•Upstream barriers remain in place to have long-term advantages	•Not effective against downstream migration of all fish •Possibly not effective under flood conditions •High cost
Explosives	•Low cost •Effective in small areas	•Generally cannot eliminate entire populations •Could impact dam integrity •Hazardous to humans and nontarget organisms •Resistant species

tuation techniques, (6) streamflow augmentation techniques (e.g., create water temperatures or current conditions that negatively impact the species to be reduced or that favor species that will prey on the undesirable species), (7) fish barriers (i.e., protect against entry by undesirable fish), and (8) explosives for flowing waters and impoundments. Advantages and limitations of these techniques are presented in Table 1.1.

1.2 SAMPLING FISH POPULATIONS (Adapted from Davies and Shelton 1983)

The main reason to sample fish populations is to determine the structure and dynamics of sport fish populations. One technique for sampling fish populations is the application of rotenone to sections of a water body. Fisheries managers gain biological information concerning the entire fish community by blocking off and treating representative areas within a water body with rotenone.

Rotenone sampling can help estimate (1) standing crop, (2) biomass, (3) total number of species present in a body of water, (4) relative abundance of year-classes, (5) size distribution of fish species or populations, (6) age and growth, and (7) predator–prey dynamics. Rotenone application can also be used to (1) investigate the relationships between fish communities and habitat characteristics, (2) build models for fisheries management, and (3) obtain information unavailable from other methods of sampling.

Table 1.2 Advantages and limitations of sampling techniques.

Type of sampling	Advantages	Limitations
Rotenone	•Sample entire fish community •Low species and size selectivity •Not dependent on diel behavior changes •Effective for shallow-water sampling •Effective in large streams where other techniques do not work •Standard area calculations of standing crop and biomass	•Most fish will be sacrificed •Public relations problems •Sample size problems •Limited by water temperature, water depth, wind, vegetation, wave action, currents •Estimates can be limited by predation: birds, alligators, turtles, other fish •May be biased toward area that can be sampled
Electrofishing	•Most sampled fish can be released live after biological information is obtained •Certain species are extremely vulnerable to sampling with electricity •Effective in shallow water •Can do many samples in a short period of time •Can target species or size-classes •Can sample diet of fish species	•Limited by water depth, flow, water clarity, temperature, conductivity, and weather •May be limited by diel activities of certain species •Not effective on scaled fish at depths of >2 m •Can be species and size selective •Escapement of fish at edge of electric field •Can be hazardous to electrofishers •Post-sampling mortalities can be high in fish
Nets	•Some sampled fish can be released live after biological information is obtained •Can get total area standing crop estimates of certain species with certain gears (e.g., seines) •Can target specific species or size-classes •Passive nets can fish for extended periods of time	•Limited by water depth, substrate, current, weather, vegetation, gradient, etc. •Species and size selective •Net avoidance or escapement •Can be time intensive •Estimates can be limited by predation: birds, alligators, turtles, other fish •Can become clogged with debris
Explosives	•Low cost	•Could impact dam integrity •Hazardous to humans and nontarget organisms
Underwater observations and sampling	•Can observe fish without capture •Can determine behavior, habitat preferences, abundance, etc. •Can observe fish community as a whole in area sampled	•Limited by water clarity, habitat, depth, etc. •Avoidance or changes in behavior in areas occupied by observers •Can be species and size selective •Limited by amount of area to be sampled

Table 1.2 continued.

Type of sampling	Advantages	Limitations
Hook and line	•Fish can be sampled and released live after biological information is obtained •Can target specific species and size-classes •Can get volunteers to supply fish for data collection	•Species and size specific •Sample size may be a problem •Weather dependent •Dependent on fishes' strike reaction •Gear bias
Hydroacoustics	•Sample large volumes and areas of water rapidly •Data can be processed rapidly by computer •Less expensive and labor intensive	•Need calm water •Need concurrent netting •No data from surface to 2 m deep •Not effective in shallow lakes or depths <2 m •Expensive equipment •Fish size limitations

1.2.1 Sampling techniques

Other sampling techniques include (1) electrofishing, (2) nets (e.g., gill, trammel, hoop, trap, cast, lift, seine, trawl), (3) explosives, (4) underwater observations and sampling using scuba gear or snorkeling, (5) hook and line (e.g., trotlines, rod and reels, limb lines), and (6) hydroacoustics. Table 1.2 presents advantages and disadvantages of these methods.

1.3 CURRENT LEGAL AND ENVIRONMENTAL CONSTRAINTS

The U.S. Environmental Protection Agency (USEPA) has approved the use of rotenone formulations to control and sample fish populations in lakes, ponds, reservoirs, and streams. Rotenone also has a long history of use as a general insecticide for treatment of agricultural crops and livestock. The powdered and liquid formulations of rotenone are designated by USEPA as restricted-use pesticides due to their aquatic toxicity and potentially adverse effects on humans through inhalation. The powdered rotenone formulations are further designated for restricted use due to their acute oral toxicity. See samples of rotenone labels in this manual for details.

Only certified pesticide applicators employed by state, federal, and provincial natural resources agencies or private persons with permits issued by these agencies after proper consultation and licensing can use rotenone formulations. Rates of application must correspond to ranges specified on the label. Other restrictions on rotenone use include (1) proper disposal of unused products and containers, (2) prohibited use of fish killed by rotenone application for food or feed, (3) prohibited use of water treated with rotenone to irrigate crops, and (4) prohibited release of treated water within one-half mile upstream of potable water or irrigation water intakes in a standing body of water such as a lake, pond, or reservoir. Prohibited uses of rotenone-treated water refer only to water intakes where water is being taken directly from the treated water body and not where alternative water supplies are provided to municipalities during and after the treatment until rotenone has dissipated. Waters or fish containing rotenone residues cannot be consumed because USEPA has not established residue tolerances (i.e., residue concentrations in water or food consumed by humans that are permitted and considered safe by regulatory agencies), not because there is a threat of toxicity to consumers at the rotenone concentrations present in fish flesh or water after typical applications. The USEPA issued a statement concluding that rotenone use for fish control does not present a risk of unreasonably adverse effects to humans and the environment. Swimming is not allowed in rotenone-treated water until the application has been completed and all the rotenone has been thoroughly mixed into the water. Inclusion of this reentry statement in the "Directions for Use" but not in the "Precautionary Statements" on the label indicates that USEPA is not concerned about human exposure to rotenone in the water.

The irrigation prohibition is intended to prevent irrigation of crops with water that contains detectable levels of rotenone. Several states have adopted the policy that rotenone-treated water will not be allowed to enter irrigation systems or potable water supplies until rotenone concentrations are undetectable. The policies of many state natural resources agencies require notification of all municipal and private potable surface water suppliers within the treatment area to advise switching to alternate water supplies. An alternative is to apply potassium permanganate during the application period to remove rotenone (Horton 1997). The USEPA has not established a tolerance for rotenone in potable water; however, the estimated safe level (4–14 μg/L) in potable water approximates the analytical detection limit of 5 μg/L rotenone (MDNR 1993). In Canada, provincial procedures and policies vary from requiring a permit to dealing with applications on a case-by-case basis. British Columbia requires a permit from Agriculture Canada, which is the final authority in pesticide registration (see Section 1.5.3). The British Columbia procedures manual mandates provision of an alternate water supply to residents who use a treated body of water as a primary source of potable water (Province of British Columbia 1993).

If a treatment project is supported by federal or state funds, requires permits, or impacts lands in the United States, an environmental assessment that describes impacts and potential mitigation is generally prepared by the affected agency.

The AFS Rotenone Task Force strongly recommends that fisheries managers carefully read and follow all the label instructions in all treatment projects. Treatments involving irrigation and potable water supplies require careful effort to determine appropriate application methods if detailed instructions are not delineated on the label.

The AFS Rotenone Task Force recommends that public access be restricted in those areas where applications occur. During the application of rotenone, public use of treated waters should be discouraged when practical to ensure the protection of the public from injury by equipment or activities and protection of personnel from harm associated with interaction with the public.

1.4 PUBLIC CONCERNS AND PUBLIC INFORMATION

The AFS Rotenone Task Force recommends close interaction between the public and the fish and wildlife agency using rotenone to ensure that public concerns are adequately addressed (see Section 2). Over the past several years, rotenone use has been temporarily prohibited or limited in several states and provinces (e.g., California and Michigan) as a result of the actions by different activist groups. Killing fish by any means is increasingly a concern to certain environmental and animal rights groups. Future uses of rotenone, even for small projects, are now threatened in several states and provinces. Often, small projects generate the greatest controversy.

Recent concerns expressed by special interest groups include (1) hazards from the inert ingredients in the rotenone formulation, (2) public health impacts, (3) impacts of other chemicals in the formulations including synergists and solvents, (4) contamination of surface and ground waters with rotenone and other chemicals, and (5) impacts on nontarget aquatic organisms, birds, and mammals (Finlayson and Schnick, unpublished). The product label does not address these specific issues because some issues are emerging and much of this information is subject to need based on site-specific environmental, sociological, biological, and economic concerns and considerations. Rotenone can only be applied by licensed pesticide applicators, some of whom receive little or no training related to aquatic applications. Many of these issues are addressed in Section 5.

A public information program is crucial to educate the public on the benefits and impacts of rotenone use; however, dispelling fears may not always be possible. As more demands are placed on the continent's bodies of water and the public becomes more environmentally aware, there will be a need to respond with information on how rotenone is being used with minimal impacts.

In a 1998 survey of natural resources agencies in the United States and Canada, many agencies reported on public concerns and expressed the need for information on the following items (in order of frequency mentioned): (1) collection and disposal of dead fish; (2) impact of rotenone and other ingredients on public health; (3) impact of rotenone and the other ingredients on surface and groundwater quality; (4) adequate public notification and education; (5) impact of rotenone on animal welfare—fish; (6) impact of rotenone on animal welfare—wildlife; (7) impact of rotenone on invertebrates; (8) rotenone residues in fish; (9) liability and property damage; and (10) impact of rotenone and other ingredients on air quality (see Appendix A).

The most important issues facing natural resources agencies using rotenone to manage or sample fish populations are public acceptance and understanding. Public acceptance issues include use of chemicals in water and killing fish. Specific issues related to the lack of public understanding include a lack of familiarity with management decisions, project purposes, and beneficial uses of rotenone to restore or sample fish populations (see Appendix A).

Colleges and universities that offer fisheries management courses are also interested in obtaining more information on proper rotenone use. Forty-one schools responded to a survey of 75 schools listed in *Fisheries Programs and Related Courses at North American Colleges and Universities* (available from AFS); only one did not provide a fisheries management course or teach about rotenone. When asked whether their courses covered use of fish chemicals, the schools responded as follows: (1) 21 covered the subject in detail; (2) 11 discussed it briefly as part of one lecture; (3) 3 reported they barely mentioned the subject; and (4) 5 did not mention the use of fish chemicals. When asked how much time was devoted to the safe use of rotenone, schools responded in the following manner: (1) 4 schools indicated that they devoted up to two lectures as well as field trips to watch the chemical applied; (2) 7 gave the subject one lecture (about 1 h); (3) 10 spent one-half a lecture on the topic; (4) 8 gave the subject less than one-half a lecture; and (5) 12 indicated that the subject was not covered.

1.5 REGISTRATION AND REREGISTRATION STATUS

Legal use of piscicides (i.e., any pesticides) requires their registration in the United States or Canada. The USEPA and the Canadian Pest Management Regulatory Agency (PMRA) currently have jurisdiction and authority over the registrations of piscicides in their respective countries. State and provincial governments may have additional registration or use requirements. Check with these authorities before any application.

1.5.1 U.S. Environmental Protection Agency

The USEPA and its predecessor, the Pesticide Regulation Division of the U.S. Department of Agriculture, intensified the regulation of piscicides in the last 35 years. The U.S. Congress passed major amendments to the Federal Insecticide, Fungicide, and Rodenticide Act in 1972 and 1988 that had dramatic impacts on the development and registration of all pesticides, especially those used on minor crops (e.g., fish populations) that have low sales volume compared to registration cost. The 1972 amendment known as the Federal Environmental Pesticide Control Act of 1972 (or FEPCA) required that all pesticides be registered for each use and reclassified and that USEPA develop regulations including certification requirements for pesticide applicators (USEPA 1973). These new requirements for registration were expensive. The annual use of aquatic pesticides is limited and chemical companies have minimal interest in registering aquatic pesticides because of low economic returns for the specialized use. As a result, registrants often offer no support or refuse to give the federal government authorization to proceed with registration of a pesticide for aquatic use even though the product is effective. The reregistration of rotenone for piscicidal use depended upon help from interested public sources.

1.5.2 U.S. Fish and Wildlife Service and the U.S. Geological Survey

Currently, only four piscicides are registered by USEPA or are in the process of reregistration (Joint Subcommittee on Aquaculture 1994): rotenone, antimycin, Lamprecid®, and Bayluscide®. The registrations and their maintenance have been possible only through federal funding by the U.S. Fish and Wildlife Service (USFWS) and U.S. Geological Survey's (USGS) Biological Resources Division (formerly the research division of USFWS).

Rotenone was first registered in 1947 in the United States by S. B. Penick & Company (now AgrEvo Environmental Health, Inc., Montvale, New Jersey). The USEPA challenged the reregistration in 1976 when it became aware of a study alleging that rotenone might be a carcinogen. While reclassifying pesticides in response to FEPCA, the USEPA considered listing rotenone as a possible carcinogenic candidate under Rebuttable Presumption Against Registration status. The claims of carcinogenicity were proven false in September 1981, but the USEPA data requirements for the classification challenge initiated a federal–state cooperative effort to reregister rotenone. The USFWS joined with the International Association of Fish and Wildlife Agencies to fund the research effort at the Upper Midwest Environmental Sciences Center (UMESC, formerly the National Fisheries Research Laboratory) at La Crosse, Wisconsin. Funding included US$1.9 million from Federal Aid in Sport Fish Restoration administrative monies contributed from 1978 to 1986 and $1.1 million from USFWS Division of Fisheries and Wetlands Research base funds committed from 1975 to 1986. Results of the research showed that rotenone is safe to the environment and humans when used according to label conditions (Sousa et al. 1987b).

On 1 June 1989, UMESC submitted a 17-volume rebuttal that challenged USEPA's Registration Standard on Rotenone because it did not include all data submitted or developed by UMESC and did not consider previous reviews by USEPA (R. A. Schnick, submitted to the Office of Research Support for the USEPA, unpublished, 1989). Since 1989, AgrEvo Environmental Health, Inc. has led the rotenone reregistration effort with assistance from UMESC. Several rotenone registrants (AgrEvo Environmental Health, Inc., Foreign Domestic Chemicals Corporation, and Prentiss Inc.) have formed a Rotenone Task Force that currently only supports rotenone's use as a piscicide. Organic farmers have formed a task force with other registrants under the National Research Support Program Number Seven to maintain the registrations of rotenone products for certain insecticidal uses, not piscicidal uses. These insecticidal registrations include uses only on vegetables and flowers, not on animals. The registrants do not expect completion of rotenone reregistration for any use until 2002 (J. Conti, AgrEvo Environmental Health, Inc., personal communication, 1998).

1.5.3 Canadian Pest Management Regulatory Agency

In Canada, any product with pesticidal claims must be registered under the Pest Control Products Act (PCPA) and under regulations administered by the Canadian Pest Management Regulatory Agency (PMRA). This agency was established in April 1995 to administer the PCPA under Health Canada rather than the Ministry of Agriculture and to support the competitiveness of agriculture, forestry, other resource sectors, and manufacturing. The PMRA assesses the toxicity, persistence, and bioaccumulation of each pesticide product, while addressing potential for human exposure and possible health hazards. The provinces and territories regulate the sale, use, and disposal of pesticides within their jurisdictions.

The "Registration Handbook for Pest Control Products Under the Pest Control Products Act and Regulations" provides information on the pesticide registration process. Under these guidelines, rotenone is considered mainly as a restricted pest control product because of its intended use in aquatic areas considered environmentally sensitive. Currently, five restricted, one commercial, and three manufacturing rotenone products are registered and available in Canada. The commercial product is intended for use by operators engaged in farming or commercial pest control operations. The three manufacturing products can also be sold for end-use purposes to certified applicators. All rotenone products can be used in lakes and flowing waters to control bullheads (*Ameiurus* spp.), carp (Cyprinidae), chubs (Cyprinidae), freshwater catfish (*Ictalurus* spp.), suckers (Catostomidae), and fish in general (F. Santagati, Pest Management Regulatory Agency, personal communication, 1999).

Rotenone was first used in Canada as a piscicide in 1937 and registered thereafter. It was also registered as an insecticide for use on a variety of vegetables, flowers, birds, companion animals, and livestock. Rotenone products are currently registered for piscicidal and insecticidal uses, but rotenone is included in the list of active ingredients for reevaluation in Canada. However, with a reevaluation target date of 2005 or 2006, rotenone is not a priority pesticide for the reevaluation program (Santagati, personal communication).

1.5.4 Current registrants of rotenone (Joint Subcommittee on Aquaculture 1994; Conti, personal communication; Santagati, personal communication)

AgrEvo Environmental Health, Inc., 95 Chestnut Ridge Road, Montvale, NJ 07645; 201-307-9700 (AgrEvo's registered products are no longer available.)
- Noxfish® Fish Toxicant (U.S. Registration No. 432-172; Canadian Registration No. 14,558)
- Nusyn-Noxfish® Fish Toxicant (U.S. Registration No. 432-550; Canadian Registration No. 19,985)
- Pro-Noxfish® Dust Fish Toxicant (U.S. Registration No. 432-829)

C. J. Martin Company, PO Box 630009, Nacogdoches, TX 75963; 409-564-3711
- Martin's Rotenone Powder® (U.S. Registration No. 299-227)

Drexel Chemical Company, 1700 Channel Avenue, Box 13327, Memphis, TN 38113-0327; 901-774-4370
- Pearson's 5% Rotenone Wettable Powder® (U.S. Registration No. 19713-316)

Foreign Domestic Chemicals Corporation, 3 Post Road, Oakland, NJ 07436; 201-651-9700
AK Product of Peru Cube Powder® (U.S. Registration No. 6458-6)
Rotenone Powder, Technical® (Canadian Registration No. 21,423)
Prentiss Inc., C. B. 2000, Floral Park, NY 11001; 516-326-1919
Prentox® Prenfish™ Common Carp Management Bait (U.S. Registration No. 655-803)
Prentox® Prenfish™ Grass Carp Management Bait (U.S. Registration No. 655-795)
Prentox® Prenfish Toxicant (U.S. Registration No. 655-422)
Prentox® Rotenone Fish Toxicant Powder (U.S. Registration No. 655-691)
Prentox® Synpren-Fish Toxicant (U.S. Registration No. 655-421)
Sureco, Inc., 9555 James Avenue South, Suite 200, Bloomington, MN 55431
Fish-Tox-5® (U.S. Registration No. 769-309)
Tifa Limited, 50 Division Avenue, Millington, NJ 07946; 908-647-4570; distributed in Canada by Dalton Chemical Laboratories Inc., Room 119, Farquharson Building, 4700 Keele Street, Toronto, Ontario M3J 1P3, Canada; 416-736-5394
Chem Fish Regular® (U.S. Registration No.1439-157; Canadian Registration No. 22,445)
Chem Fish Synergized® (U.S. Registration No.1439-159; Canadian Registration No. 22,447)
Powdered Cubé Root Manufacturing Concentrate® (Canadian Registration No. 22,444)
Chem-Fish Special® (Canadian Registration No. 22,446)
Zeneca Agro, a business of Zeneca Corporation, 250-3115 12th Street N.E., Calgary, Alberta T2E 7J2, Canada; 403-219-5400
Rotenone Fish Poison Wettable Powder® (Canadian Registration No. 16,580)

1.6 Terminology

The words "must," "should," "may," "can," and "might" have very specific meanings in this manual:

- "Must" is used to express an absolute requirement, that is, to state that the guidelines are designed to satisfy the specified condition. "Must" is only used in conjunction with factors that directly relate to the legality or acceptability of specific recommendations (i.e., a requirement on the label of a pesticide product).
- "Should" is used to state that the specified condition is recommended and ought to be met, if possible. Terms such as "is desirable," "is often desirable," and "might be desirable" are used in connection with less important factors.
- "May" is used to mean "is (are) allowed to."
- "Can" is used to mean "is (are) able to."
- "Might" is used to mean "could possibly." "Might" is never used as a synonym for either "may" or "can."

1.7 Glossary of terms

bioaccumulation The potential for a substance to accumulate in living biological tissue

control Reduction of fish populations or fish species

dispersant A substance that assists in spreading another substance

emulsifier Generally a petroleum-based substance in water; a substance used to stabilize the suspension of one liquid in another

eradication Elimination of whole fish populations or fish species from distinct habitats or bodies of water

half-life The time period in which one-half of an amount of substance degrades

hydrolysis The decomposition of a substance through reaction with water

LD50 A statistically derived estimate of a concentration of a substance that would cause 50% mortality to the test population under specified conditions

oxidation The decomposition of a substance by uniting with oxygen

pesticide Any substance or mixture of substances intended for preventing, destroying, repelling, or mitigating any pest

photolysis The decomposition of a substance caused by exposure to light

piscicide Chemical toxic to fish that is used to control, eradicate, or sample fish populations

restricted-use pesticide Pesticide restricted for a specific reason usually related to safety; to be used only under the direct supervision of a certified pesticide applicator

tolerances Residue concentrations of a chemical that are permitted by regulatory agencies in water or food consumed by humans

undesirable fish Species of fish designated by fisheries managers as undesirable in certain bodies of waters

volatile organic compounds Mainly petroleum-based substances that vaporize freely into air

Appendix A

Rotenone Use in North America (1988–1997)*

by

William McClay

Lake Management Specialist, Retired

Fisheries Division, Michigan Department of Natural Resources

* This appendix previously appeared as an article in *Fisheries* 25(5):15–21.

INTRODUCTION

Fisheries managers rely on a wide variety of tools for the management and assessment of fish populations to maintain diverse and productive aquatic ecosystems and high quality recreational fisheries. One of the most valuable tools is the piscicide rotenone, which was first used in the United States in 1934 in Michigan (Ball 1948; Lennon et al. 1970; Cumming 1975) and in Canada in 1937 (M'Gonigle and Smith 1938). The use of rotenone as a fisheries management tool is taught in at least 38 of 75 North American colleges and universities that teach fisheries programs and related courses (G. Tichacek, retired, Illinois Department of Conservation, personal communication). Techniques for the use of rotenone to sample fish communities and for reclamation and fish control activities are covered extensively by Bettoli and Maceina (1996).

Important uses of rotenone in fisheries management include:

- control of undesirable fish
- eradication of harmful exotic fish
- eradication of fish in rearing facilities and ponds to eliminate competing species
- quantification of populations
- treatment of drainages prior to impoundment
- eradication of fish to control disease
- restoration of threatened or endangered species

The application of a piscicide is the only method other than complete dewatering that will extirpate entire populations of fishes. Complete elimination of fish is often needed to accomplish the critical fish management activities of removing predatory exotic species, restoring threatened and endangered species, and controlling fish diseases. Rotenone is the only sampling method that provides for an accurate estimation of standing crop of diverse fish communities.

Despite the importance of rotenone in fisheries management, its continued availability and use are uncertain. Most rotenone treatments have occurred without incident; however, putting any chemical into water, especially one that kills fish, can create controversy.

A small number of treatments have resulted in public controversy. Incidents in California, Colorado, Michigan, and Minnesota resulted in adverse public reaction and negative publicity in the news media. Some of these incidents could possibly have been avoided if the responsible agency had (a) garnered more public input and support prior to treatment and not been in an adversarial role with local communities, (b) done a better job of implementing the treatment with appropriate procedures and qualified personnel, or (c) provided better technical, administrative, legal, and political support. Public relations issues included fish mortalities downstream of the application site and persistence of treatment chemicals in water and air. As a result, the use of rotenone was temporarily prohibited in one state (Michigan) and has been limited in several others.

The use of rotenone is increasingly a concern to environmental and animal rights groups, and the future use of rotenone, even for small projects, has been threatened in several states, most notably New York and California. As more demands are placed on the continent's water bodies and the public becomes more environmentally aware, we must respond with guidelines to use rotenone prudently with minimal impacts and controversy.

In 1993, the Task Force on Fishery Chemicals of the American Fisheries Society submitted a proposal to develop and implement a Rotenone Stewardship Program for fisheries management using U.S. Fish and Wildlife Service Federal Aid Administrative Funds. The proposal was accepted for funding in 1997. The first task was to conduct a survey of current uses, issues, and restrictions so the stewardship plan would reflect current knowledge and concerns.

METHODS

A detailed questionnaire was sent to fisheries management agencies in all of the provinces of Canada, the states of the United States, the District of Columbia, and to regional U.S. Fish and Wildlife Service offices. The survey was not sent to other federal agencies, universities, private consultants, or private individuals that use approximately 10% of the total sales of raw material (R. Fisher, AgrEvo Environmental Health, Inc., personal communication).

Agencies were asked to report rotenone usage (liquid or powder formulation) for the 10-year period of 1988–1997 by type of water body (standing or flowing). They also were asked to identify issues experienced when using rotenone, and indicate what type of information and guidance they needed in a handbook of administrative and technical procedures.

The survey requested information on the weight of powder and volume of rotenone formulations used. It was difficult to compare use among the different formulations (5% powder, 5% liquid, and 2.5% syngerized liquid) because of the different percentages of rotenone in each formulation. Therefore total quantities of rotenone in the various formulations were converted to kilograms of active rotenone used. Thus, in this appendix, all references to kg of rotenone refer to kg of active ingredient. This allows comparisons of the quantities of rotenone used between the two 5-year periods of the survey (1988–1992 and 1993–1997), among purposes of treatment, and between water types (static and flowing waters), regardless of formulation. The conversion assumes rotenone is 5% by weight in all liquid and powder formulations. Liquid formulations contain either 5% rotenone by weight or 2.5% rotenone by weight with a 2.5% synergist by weight. Powder is generally sold on a 5% rotenone by weight basis. The synergized formulations are used as if they were 5% weight formulations (i.e., the treatment rate is not generally doubled because of the reduced rotenone content). Use data statistics were analyzed with and without the 1990 data for Strawberry Reservoir (Utah) because of the effect this treatment had in skewing the data. This one treatment required 20,000 kg of powdered rotenone, representing 43% of the powder used during the 10 years covered by the survey.

A summary of the results from the survey and their significance are discussed below.

RESULTS

A total of 95 questionnaires were sent to 68 United States and 20 Canadian agencies, the District of Columbia, and 7 U.S. Fish and Wildlife Service offices. A total of 78 (82%) responses were received. Responses were received from 55 state agencies (80%) representing 48 of the 50 states and 15 Canadian agencies (75%) representing 11 of the 12 provinces and territories. Responses were also received from the District of Columbia and seven U.S. Fish and Wildlife Service offices. Several states and provinces had more than one agency respond because of divided management responsibilities. Responses were not received from Arizona, Colorado, or Saskatchewan. Information about the Northwest Territories was included in the response from Manitoba.

Most agencies indicated the data on the quantities of rotenone used for various purposes in standing and flowing waters were reliable and based on verifiable records. However, a few agencies indicated some of their historical records did not allow them to differentiate quantities used for various purposes. Therefore, some of the quantities reported for specific purposes were estimated by the agency.

Scope of use

Of the 78 responding agencies, 48 (62%) reported using rotenone in the last 10 years (1988–1997). Rotenone was used in 37 states (77%) and 5 provinces or territories (42%) during the survey period. Thirty-three of the states and 4 provinces used rotenone as recently as 1997.

Of the 29 responding agencies who did not use rotenone:

- eight were responsible for managing marine environments and indicated rotenone was not effective in systems with tidal and wave currents.
- five agencies indicated they used rotenone 15–20 years ago and had no need to do so now.
- nine agencies did not provide reasons for not using rotenone.
- seven agencies indicated they stopped using rotenone due to in-house policies, administrative requirements, or regulations (five), expense (one), or because of environmental concern (one).

Quantities of rotenone used

During the 10-year period, a total of 94,739 kg of rotenone were used (Figure 1). However, 20,695 kg (22%) were used on one project (Strawberry Reservoir, Utah) in 1990 (Figure 1). The treatment of Strawberry Reservoir accounted for 1.4% of the liquid and 42.6% of the powder used during the survey period. Rotenone use declined 57% from the first (1988–1992) to the second (1993–1997) 5-year period of the survey when the amount used in Strawberry Reservoir is included and declined 38% when it is excluded (Figure 1).

The preferred formulation of rotenone used appears to have changed between the two 5-year periods of the survey (Figure 2). Agencies now appear to be placing greater emphasis on the use of powder where practical. The amount of liquid rotenone used declined 65% (35,406 kg to 12,405 kg) from the first to the second 5-year period. The influence of the Strawberry Reservoir treatment on liquid use was minimal.

The amount of powdered rotenone used declined 49% (31,053 kg to 15,875 kg) from the first to the second 5-year period (Figure 2). However this decline is not a true representation of use because the data is skewed by the Strawberry Reservoir treatment in 1990 (in the first 5-year period). This treatment required 20,000 kg of the 31,053 kg of powder used (64%) in the first 5-year period. When the Strawberry Reservoir data is excluded, powdered rotenone use actually increased 44% (11,053 kg to 15,875 kg) from the first to the second 5-year period (Figure 2).

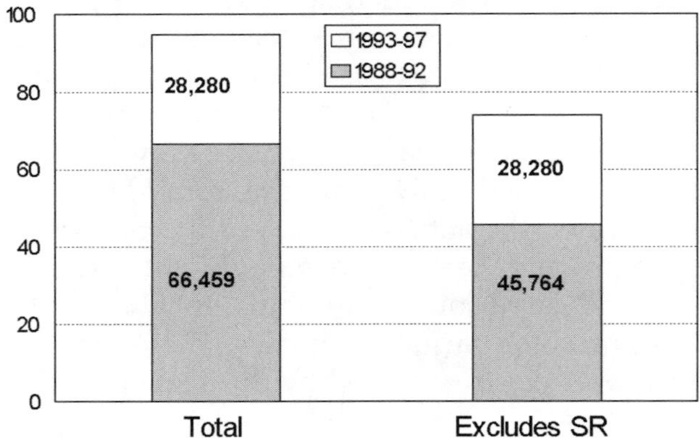

Figure 1 Compares quantities of rotenone (kg active ingredient from all formulations) used in the United States and Canada during the two 5-year periods of 1988–1992 and 1993–1997. Shown are quantities including and excluding rotenone used in the large 1990 Strawberry Reservoir, Utah (SR), treatment.

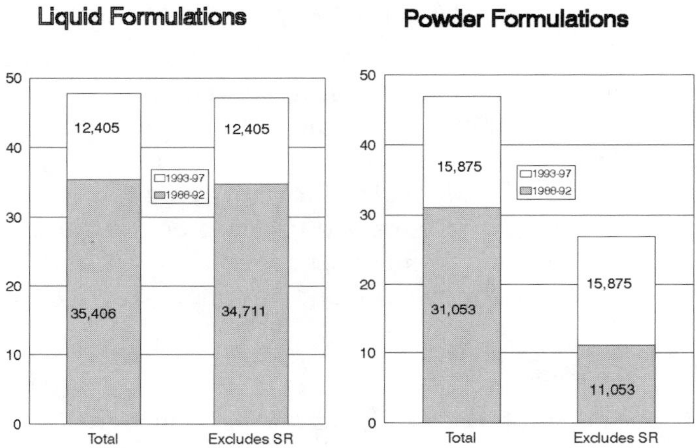

Figure 2 Compares quantities of rotenone (kg active ingredient) used from liquid and powder formulations in the United States and Canada during the two 5-year periods of 1988–1992 and 1993–1997. Shown are quantities including and excluding rotenone used in the large 1990 Strawberry Reservoir, Utah (SR), treatment.

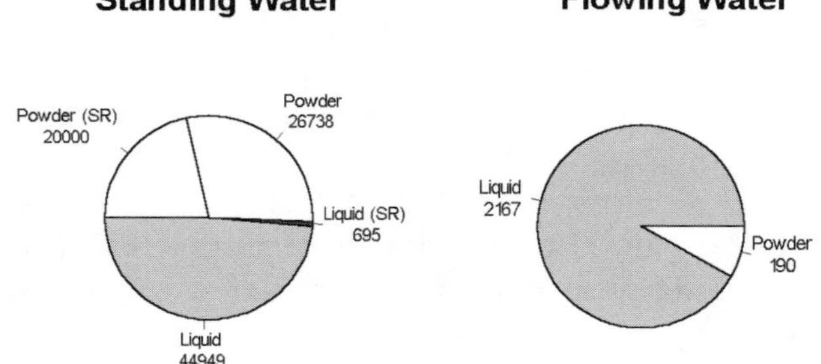

Figure 3 Compares quantities of rotenone (kg active ingredient) used from liquid and powder formulations in standing and flowing waters of the United States and Canada during the period of 1988–1997. Shown are quantities including and excluding rotenone used in the 1990 Strawberry Reservoir, Utah (SR), treatment.

Most of the rotenone (97.5%) used during the survey period was applied to standing water (Figure 3). Of the 92,382 kg of rotenone applied, nearly equal amounts came from liquid (45,644 kg) and powder (46,738 kg) formulations (Figure 3). In flowing waters, 92% (2,167 kg) of the rotenone came from liquid formulations. This difference reflects the inability to effectively use powdered rotenone in flowing waters.

Uses of rotenone

Manipulation of fish communities to maintain sport fisheries was the most common reason for using rotenone (Figure 4). This type of treatment accounted for 42% of the waters treated (2,050 treatments) using 72% (68,944 kg) of the rotenone. Of the 1,838 km of streams treated, 38% (697 km) were treated to maintain sport fisheries, and of the approximately 400 hm^3 of standing water treated, 40% (160 hm^3) were treated to maintain sport fisheries (Figure 5).

Quantification of fish populations (sampling) was the second most common purpose (Figure 4). This accounted for 31% (1,482 treatments) of waters treated, illustrating the importance many agencies place on this sampling technique. Fourteen of the 37 states (38%) indicated they used rotenone for this purpose. This accounted for 4% (84 km) of the flowing water treated and 6% (23 hm^3) of the standing water treated (Figure 5). Although a significant use of rotenone in terms of the number of waters treated, the volume of water treated and quantity of rotenone used (2,114 kg) were minimal, indicating the treatments were small.

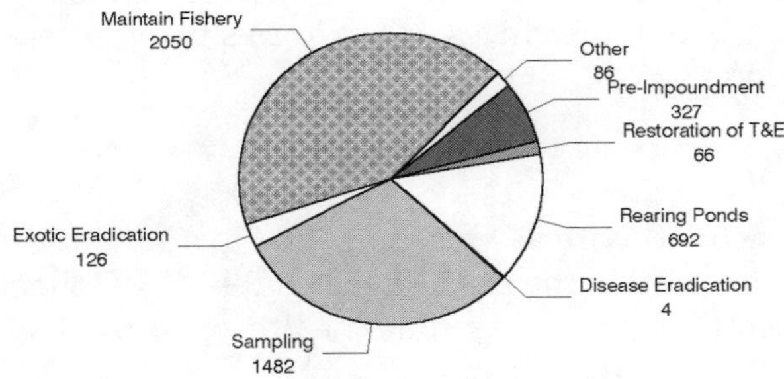

Figure 4 Compares the number of rotenone treatments by objective conducted in the United States and Canada during the period of 1988–1997.

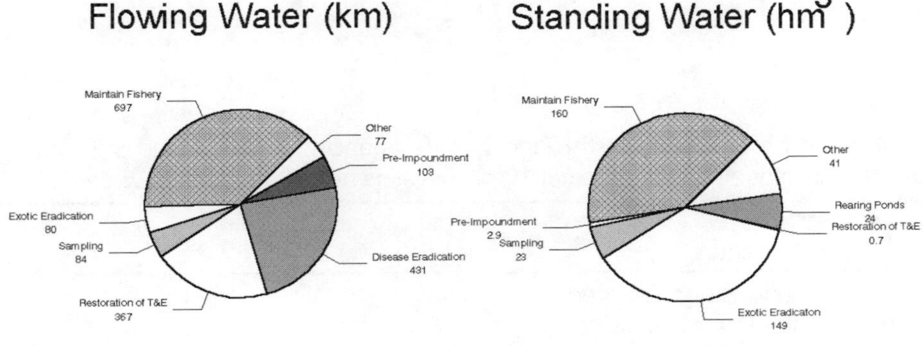

Figure 5 Compares the length (km x 1,000) of flowing water and volume (hm^3 x 1,000) of standing water treated with rotenone by objective in the United States and Canada during the period of 1988–1997.

The treatment of rearing facilities or rearing ponds ranked third in terms of the number of waters treated (Figure 4). Rearing facilities and rearing ponds represented 14% (692 treatments) of the total number of waters. Many states did not provide actual numbers even though estimates of volume treated and gallons used were provided. Twelve of the 31 states (39%) indicated they used rotenone for this purpose. The volume of water treated was small (24 hm^3) compared to other purposes for treating standing water (Figure 5).

Treatments aimed at the eradication of exotic fish ranked fifth (126 treatments) in terms of the number of waters treated (Figure 4), but ranked second (149 hm^3) in terms of the total amount of standing water treated (Figure 5) and used 18% (17,219 kg) of the rotenone.

Rotenone treatment procedures

Agencies were asked to respond with a "yes" or a "no" to questions on whether they used specific treatment procedures (Table 1). The majority indicated that permission was required to use rotenone, that they detoxified with potassium permanganate, and that chemical monitoring was not performed.

Regulation of rotenone

Rotenone is not regulated by the agencies that use it. Of the 37 states using piscicides, the majority are regulated by a state department of agriculture or by a state environmental agency. All five Canadian provinces or territories using rotenone are regulated by an environmental agency.

Issues related to rotenone

Agencies were asked to respond with a "yes" or a "no" to a series of questions relating to specific issues that have arisen in the past 10 years (Table 2). Collection and disposal of dead fish and impacts of rotenone on public health were the two most numerous issues mentioned.

Other less frequently listed issues included: (a) killing of game fish and public perception about piscicide use; (b) U.S. Forest Service ban on the use of piscicides; (c) cattle grazing near treated water; (d) registration; and (e) loss of use as a survey technique.

Table 1 Treatment procedures that fish and wildlife agencies in the United States and Canada utilize for rotenone. Not all responses total 100% because not all agencies answered all questions.

Treatment procedure	Number of agencies	Yes	No
Government permit/permission required	48	81%	19%
Detoxify with potassium permanganate	46	72%	21%
Environmental impact analyses/assessment	48	48%	52%
Effectiveness and impacts monitored:			
With bioassay	48	42%	58%
With water samples for chemical analysis	45	25%	69%

Table 2 Rotenone-related issues that fish and wildlife agencies in the United States and Canada have addressed during the 10-year period of 1988–1997. Not all responses total 100% because not all agencies answered all questions.

Issue	Number of agencies	Yes	No
Collection and disposal of dead fish	48	48%	52%
Impact of rotenone (or other ingredients) on public health	46	42%	54%
Impact of rotenone (or other ingredients) on surface or ground water quality	47	31%	67%
Adequate public notification and education	47	31%	67%
Animal welfare, fish	47	31%	67%
Animal welfare, wildlife	46	29%	67%
Impact on invertebrates	47	27%	71%
Piscicide residues in fish	46	21%	75%
Liability and property damage	46	15%	81%
Impact of rotenone (or other ingredients) on air quality	47	8%	90%
Other issues (see text)	44	13%	79%

Most important issues facing users of rotenone

Agencies using rotenone were asked to provide their view of the two most important issues facing users of rotenone. Thirty-nine of the 49 agencies responded by citing 83 different issues. Many agencies cited the same or similar issues. These issues were grouped into eight categories based on their similarities (Table 3). Agencies overwhelmingly identified public acceptance and understanding, environmental concerns, and "usability" of the product as the most important category of issues confronting them.

Public acceptance and understanding of rotenone use was the most frequently mentioned issue category. Issues most often mentioned were a lack of public knowledge and understanding of the management decisions that led to rotenone treatments, the purpose of the project, and the beneficial uses of rotenone. Also mentioned frequently was a lack of public acceptance for using chemicals in the water and for killing fish. Agencies cited complications arising when opposition groups became organized, secured financing, and mounted legal challenges.

Table 3 Rotenone-related issue categories that fish and wildlife agencies in the United States and Canada considered important, by order of frequency.

Issue category	Rank
Public acceptance or understanding	1
Environmental concerns	2
Usability of the product	3
Public health or toxicology concerns	4
Availability of the product	5
Animal rights or welfare concerns	6.5
Miscellaneous	6.5
Methods or techniques	7

Environmental concerns were the second-largest issue category. These issues had their origins from both the public sector and from those governmental agencies with management responsibilities. One frequently mentioned environmental concern focused on biodiversity-related decisions that led to (a) single-species management (e.g., waters managed for trout only), (b) management directed toward quality sport fish populations (as opposed to "nonmanaged waters"), or (c) management directed toward threatened and endangered species. Other environmental concerns focused on the impacts (real or perceived) of rotenone treatments on nontarget species such as invertebrates, mussels, amphibians, and those wildlife and domestic animals which may eat fish killed by rotenone.

"Usability" of rotenone was the third-largest issue category. Issues included restrictions on use due to labeling, legislative mandates or policy, restrictions placed on use for sampling, registration costs, and economics (agency budgets and cost effectiveness).

The fourth issue category included issues related to general toxicology, general public health, carriers and inert ingredients, and drinking water safety.

The remaining four issue categories (and major issues) were (a) availability of the product (especially relicensing and reregistration of the liquid formulation), (b) animal rights, (c) methods (stream treatments, effective detoxification, and effective use in population management and assessment), and (d) miscellaneous issues such as public distrust of state and federal agencies, documentation and control of use, lack of specific, useful, and practical information on impacts of treatments, proper training for applicators, and project goals being met for extended periods of time.

Requested information and guidance

Twenty-six of the 49 agencies made suggestions for the type of information and guidance they desired in a handbook of administrative and technical procedures. The suggestions have been grouped into the following broad categories:

Background information—(a) regulatory history; (b) national policy on the use of rotenone; (c) definitions of "restricted use chemical"; (d) literature sources; and (e) role of federal agencies in state projects.

Environmental information—(a) short-term and long-term impacts; (b) minimizing impacts on nontarget species; (c) long-term effectiveness of rotenone treatments to assist in preparing environmental documents; (d) sensitivities of various species to rotenone; and (e) water quality, persistence, and degradation.

Legal considerations—(a) who can legally purchase and use rotenone; and (b) legal considerations for fish disposal, use and disposal of rotenone and their containers, and public notification.

Management considerations—(a) when is rotenone the best management choice; (b) description of practical uses; (c) appropriate uses; and (d) alternative methods.

Public health information—(a) impact to applicators and the general public; (b) assessment for human and animal exposure; (c) cancer risk; (d) fish consumption; and (e) contact with treated water.

Public information—(a) dealing with anti-fish-treatment public; (b) public relations strategies that will garner support or minimize opposition; (c) public information on the utility of rotenone as a management tool; and (d) dealing with animal rights activists.

Technical information—(a) safe handling, storage, shelf life, and disposal; (b) effective concentrations; (c) application and detoxification procedures; (d) bioassay methods for determining toxicant level; (e) chemical analysis methods; (f) influence of environmental factors on application and detoxification effectiveness; (g) selectivity and application rates; (h) methods for reservoir fish population assessment; (i) new designs for applying powdered rotenone; (j) application rate table for specific concentrations and flows; and (k) procedures for fish disposal.

DISCUSSION

The present survey primarily targeted state and provincial fisheries agencies. The survey was also directed to the regional offices of the U.S. Fish and Wildlife Service, but not all uses were reported (i.e., use on refuges and by Cooperative Fishery Units). Additionally, numerous states reported that other federal agencies used rotenone, including the Tennessee Valley Authority, U.S. National Park Service, and Bureau of Land Management. It was also learned that some Native American tribes use rotenone on reservation lands. However, rotenone use by these agencies, consultants, and other entities was assumed to be a minor component of overall use. Sales data showed that AgrEvo Environmental Health, Inc., which captures about 85% of the piscicide market (Fisher, personal communication), sold approximately 13% of their product to federal agencies, consultants, and other groups not included in the survey. Thus, the agencies surveyed accounted for about 87% of the total piscicide market for rotenone.

The use of rotenone as a piscicide in North America is as widespread today as ever even though the quantity used has declined. In the 15 years between its first use in 1934 (in Michigan) and 1949, 34 states reported using rotenone (Solman 1950). A United Nations-sponsored report on world use of rotenone revealed that by 1970, 39 states (and two provinces) had used rotenone to reclaim waters (Lennon et al. 1970). A later survey covering the period 1970–1974, targeted exclusively at the state and territorial fisheries agencies of the United States, determined that 49 states used rotenone (S. B. Penick & Company, unpublished data, 1974).

In 1987 the rotenone supplier, AgrEvo Environmental Health, Inc., conducted a telephone survey of United States and Canadian fisheries agencies to estimate market size; however, no data on actual use were collected. A review of the data sheets from that survey revealed that 37 states and three provinces were current users of rotenone.

The present survey suggests that the quantity of rotenone used is continuing to decline, although the number of states using rotenone has changed little since 1949. Thirty-seven states (and five provinces or territories) used rotenone, which continues the 50-year trend where 35 states have historically utilized rotenone as a fisheries management tool. However, since 1974, major users of rotenone (> than 50 kg annually) have declined. Thirty-four states were major users in 1974 (S. B. Penick & Company, unpublished data, 1974), but this survey showed a decline to 25 states (and two provinces) in 1987 and a further decline to 14 states (and one province) in 1997. Quantities of rotenone used also declined during the survey period. Rotenone use declined 57% from 1988–1992 to 1993–1997. When the rotenone used in Strawberry Reservoir is excluded, use declined 38% between the two 5-year periods. Bettoli and Maceina (1996) also noted decreasing use of rotenone for sampling and reclamation of fish populations.

Despite this decline, rotenone continues to be an important management tool for most fisheries agencies in North America, and its use as a fisheries management tool continues to be taught in many colleges and universities. Nearly 95,000 kg of rotenone (liquid and powder formulations) were used during the 1988–1997 period. Managers appear to be placing greater emphasis on the use of the powder formulation, particularly for treating standing waters. This trend is probably due to the reduced cost and improved distribution techniques for the powder formulation, as well as increased environmental and public health concerns for the inert ingredients contained in liquid formulations. Although liquid formulations have been proven safe for use, some agencies have found it more difficult to plan and execute treatments using these formulations because of demands for environmental monitoring studies not normally required for projects that utilize the powder formulation.

Agencies responding to the survey provided their perception of the major issues in using rotenone. They overwhelmingly identified public acceptance and understanding of rotenone use, environmental concerns, and continued usability of the product as major issues. A common theme to many of the issues cited was the lack of public knowledge and understanding of the management decisions which led up to rotenone treatments, the purpose of the project, and the beneficial uses of rotenone. In spite of this recognition, only 48% indicated that they performed environmental impact analyses or assessments on proposed projects.

To overcome these issues, agencies must do a better job of communicating project objectives and environmental trade-offs to the public. For example, public support for renovating a fish community may be generated when managers can demonstrate that the current community is the result of human-induced perturbations and that the only alternative is complete renovation. Further, the public often does not understand that some short-term losses may be offset by long-term benefits including, but not limited to, many years of improved angling opportunity.

In response to the request by agencies for more information and guidance on the use of rotenone, this manual will assist fisheries managers by providing administrative and technical guidelines for the safe and effective use of rotenone. Emphasis is placed on planning and public involvement commensurate with the scope of the project. There are also plans for a public information program to educate the public on the benefits and risks of rotenone use. An electronic information system for fisheries biologists that will provide up-to-date information on current use restrictions, experts in the use of rotenone, important issues and solutions, and the registration status of rotenone are also under development.

ACKNOWLEDGMENTS

Brian Finlayson, chair of the AFS Fish Management Chemicals Subcommittee, and Rosalie Schnick, chair of the AFS Task Force on Fishery Chemicals, organized the Rotenone Stewardship Program, secured funding, developed the initial draft of the survey, and reviewed the manuscript. Subcommittee members Richard Cailteux, Leo Demong, Bill Horton, Gregg Tichacek, and Charles Thompson assisted in revising the survey questions and reviewing the manuscript. Fisheries scientists from 55 state agencies, 15 Canadian agencies, the District of Columbia, and 7 U.S. Fish and Wildlife Service offices provided information on their use of rotenone over a 10-year period. Ruth Fisher provided information from a 1987 marketing survey and 1986–1997 sales data from AgrEvo Environmental Health, Inc. and its predecessors. This study was funded with administrative funds from the Sport Fish Restoration Program.

REFERENCES

Ball, R. C. 1948. A summary of experiments in Michigan lakes on the elimination of fish populations with rotenone, 1934–1942. Transactions of the American Fisheries Society 75:139–146.

Bettoli, P. W., and M. J. Maceina. 1996. Sampling with toxicants. Pages 303–333 *in* B. R. Murphy and D. W. Willis, editors. Fisheries techniques, second edition. American Fisheries Society, Bethesda, Maryland.

Cumming, K. B. 1975. History of fish toxicants in the United States. Pages 5–21 *in* P. H. Eschmeyer, editor. Rehabilitation of fish populations with toxicants: a symposium. American Fisheries Society, North Central Division, Special Publication 4, Bethesda, Maryland.

Lennon, E., J. B. Hunn, R. A. Schnick, and R. M. Burress. 1970. Reclamation of ponds, lakes and streams with fish toxicants: a review. FAO (Food and Agriculture Organization of the United Nations), Fisheries Technical Paper 100.

M'Gonigle, R. H., and M. W. Smith. 1938. Cobequid hatchery - fish production in Second River and a new method of disease control. The Progressive Fish-Culturist 38:5–11.

Solman, V. E. F. 1950. History and use of fish poisons in the United States. Canadian Fish Culturist 8:3–16.

Administrative Procedures 2

A rotenone application project may have up to five stages:

(1) preliminary planning, where the project concept and alternatives are developed, public input is invited, and acceptance is encouraged;

(2) intermediate planning, incorporating an environmental analysis where the project is refined and public acceptance is encouraged;

(3) final planning and project implementation, involving management through the development of project-specific work plans;

(4) performing the treatment (see Figure 2.1); and

(5) summation and critique of the project into a final report (see Section 4.1.4).

A small treatment performed on private land or a government-owned hatchery may require little planning before implementation, while a large project involving a public water supply may require two or more years of extensive planning. The rotenone treatment should be consistent with and supported by the current Fisheries Management Plan (FMP) when applicable, which is either species specific (see Section 2.1.1.1) or water body specific (see Section 2.1.1.2). The complexity of a rotenone project depends upon social, biological, political, and physical characteristics and will dictate the degree of planning required. For example, extensive planning may not be needed for rotenone use in sampling except where downstream waters are potentially affected.

2.1 Preliminary planning

Preliminary planning is critical to the success of fish reclamation and sampling projects using rotenone. The project must be based on facts and tactics that firmly stand throughout the whole project. Key ingredients in this first stage are public input and acceptance, without which difficulties or failure are possible.

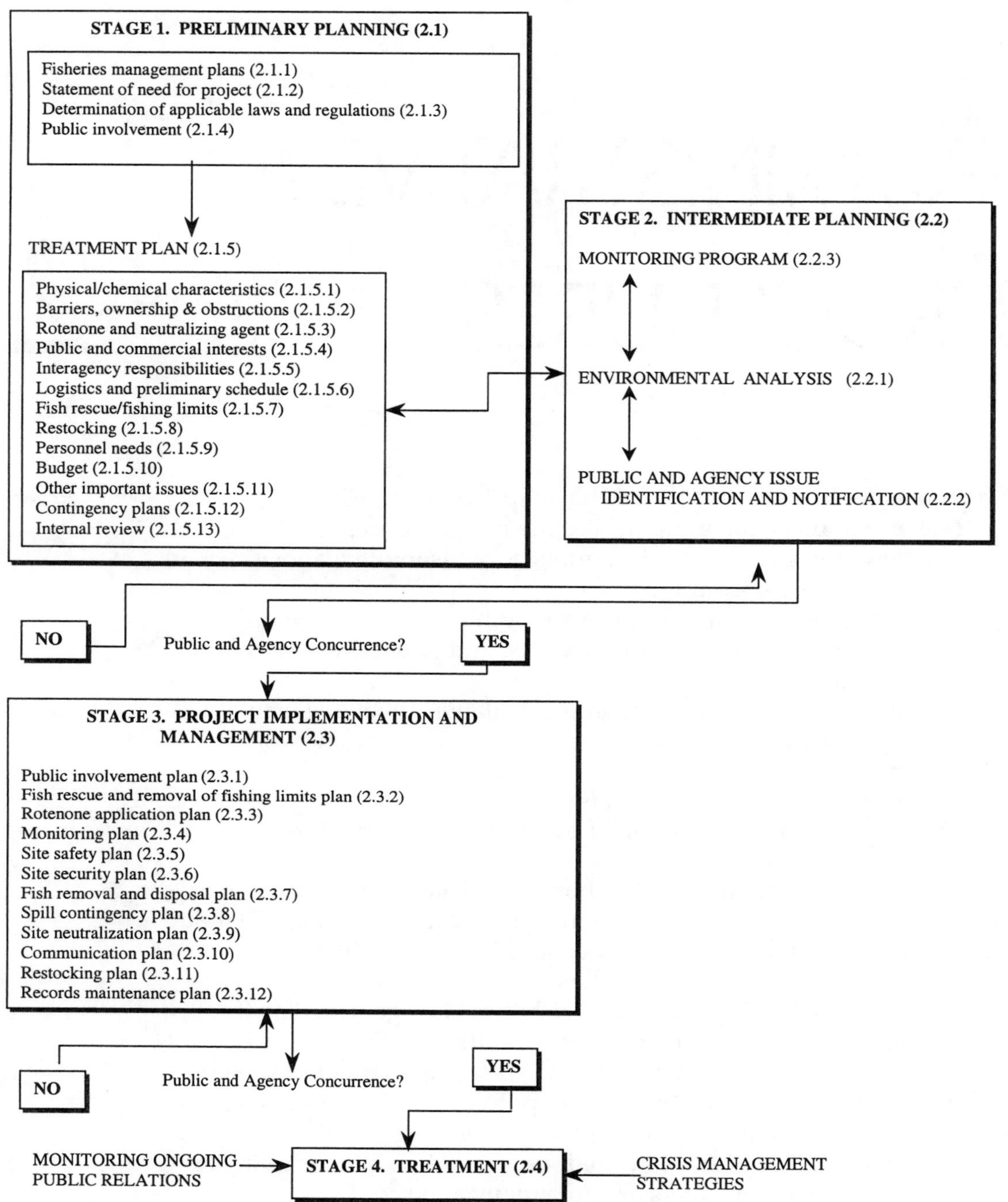

Figure 2.1 Administrative procedures for planning a rotenone treatment (section numbers in parentheses).

2.1.1 Fisheries management plans

Fisheries resources can be managed for a particular species or for a particular water body, or both. A Fisheries Management Plan (FMP) assesses the status of a specific water body or populations of specific fish species and determines the appropriate management actions necessary to maintain the desired fishery. A current FMP ensures that written goals and objectives for a specified time period are clearly defined and implemented. The type of management desired will determine which type of FMP is used. Public input during the development of the FMP is beneficial.

2.1.1.1 Species-specific fisheries management plan

A species-specific FMP minimally contains the following items: (1) goals and objectives of the FMP; (2) historic habitat range of managed species; (3) description of environmental and human problems; (4) identification and prioritization of suitable habitat locations, including threats to each habitat type; (5) plan description; and (6) time line of management measures.

2.1.1.2 Water body (location)-specific fisheries management plan

An FMP for a specific water body is similar to the species-specific plan except that this plan includes a description of the biotic diversity of the water body instead of a habitat range for the species. The water body-specific FMP should contain the following items: (1) general geographical setting of the area; (2) description of existing land management surrounding the water body; (3) water quality and development surrounding the water body (e.g., forest, residential, industrial); (4) recreational facilities and activities; (5) ownership; (6) hydrology; (7) aquatic animal assemblages and habitat types; (8) threatened and endangered species; (9) fishery description; (10) current fish management; and (11) proposed management program (objectives, direction, recommendations).

2.1.2 Statement of need for project

The reason for a project must be clearly supported by factual evidence. For example, the presence of undesirable fish as defined by the fisheries manager may suggest rotenone treatment. The objective of the project is to correct existing fishery conditions that conflict with the goals of the FMP. Project objectives might include (1) reversing unacceptable population size or growth rate of a desirable fish species, (2) eliminating undesirable species, (3) minimizing outbreaks of contagious disease, (4) reintroducing a native species into a historical range, or (5) changing the desired species composition in response to public demand. The justification could also include consideration of the current and potential demand for fishing within the water body, the need to protect nearby waters from undesirable species, or the uniqueness of a remote, pristine fishery. Projects that emphasize single species management should address issues of ecosystem diversity. Knowledge of the presence of parks or other public facilities (or possible future developments), and the proximity of population centers is useful for this process. The justification should also explain why other options would not accomplish the desired results. If an environmental assessment is planned (see Section 2.2.1), the draft plan should list all options or alternatives but should not indicate the alternative preferred by the fish and wildlife agency. Small projects that do not require an environmental assessment should include a clear justification for the decision to use rotenone.

Before proposing a project, a biological survey of fish community composition is necessary. At a minimum, sample a variety of species and age-groups to determine the presence of the most rotenone-resistant species subject to removal. The justification may contain measures of angler success and use such as creel survey information and fish stocking information. Written or oral comments solicited from the angling public can provide information about general satisfaction with a fishery.

The justification should include a description of the fish community, desired fish management objectives, life history of the target fish species, and a comparison of the available alternative control measures. The project might also have population estimation and enumeration as an objective. The appropriate uses of rotenone and alternative control measures are discussed in Sections 1.1 and 1.2.

2.1.3 Determination of applicable laws and regulations

2.1.3.1 Fish and wildlife authorities

State the legal authorization (federal, state, or provincial) for fish and wildlife agency management of aquatic resources. Documentation of this authority may prove instrumental in countering legal challenges to the project and in negotiations with other parties. These mandates usually address the conservation, maintenance, and utilization of natural resources to ensure the continued existence of all species and the maintenance of a sufficient resource to support reasonable recreational fisheries. The fish and wildlife agency may have specific powers to take any species which is (1) unduly preying upon a desirable species of bird, mammal, or fish, (2) an introduced species, or (3) harboring a highly contagious disease.

2.1.3.2 Clearances required for treatment

Determine those regulatory agencies that have overlapping jurisdictions for regulating a treatment. Agencies that regulate the following areas may require notifications, applications, approvals, and permits: (1) agriculture; (2) water use; (3) environmental protection; (4) water quality; (5) public health; and (6) land use. Determine the applicable regulations and restrictions, and obtain clearances in sufficient time before the treatment. Outside agencies may need monitoring plans and other requirements before treatment, so allow sufficient time for compliance. Resolve conflicts over regulatory and jurisdictional issues before treatment. Agreements that delineate interagency authorities, responsibilities, procedures, and time lines have been instrumental in resolving conflicts among agencies with overlapping responsibilities (see Appendix B).

2.1.4 Public involvement

2.1.4.1 Significant public issues

Determine the significant public issues that must be addressed. Although rotenone is a useful fishery management tool, its use has often resulted in considerable public controversy. This controversy stems from objections from three main groups: (1) persons who oppose changes to a perceived natural situation or oppose the use and development of so-called "monocultures" of fish; (2) persons who are alarmed by the perception of widespread application of chemicals that might be dangerous to people and the environment; and (3) persons who oppose any means of killing of fish.

In a recent survey of fish and wildlife agencies sent to all states and provinces (see Appendix A), the following issues were identified as controversial during the past 10 years, in order of frequency: (1) collection and disposal of dead fish; (2) impact of rotenone on public health; (3) impact of rotenone on surface and groundwater quality; (4) adequate public notification and education; (5) animal welfare (fish, wildlife, and invertebrates); (6) pesticide residues in fish; (7) liability and property damage; and (8) impact of rotenone on air quality. See Appendix A for a full discussion of the results of this survey.

2.1.4.2 Gaining public input and support

Obtain public input and support for the project. Anticipate the types of controversies that may arise, deal with them quickly and as openly as possible, and address them before the project is implemented. Controversy can be greatly minimized and often eliminated if the project and treatment are developed and implemented carefully and thoughtfully with public input. Reevaluate projects that do not have a high degree of public support.

The type of treatment, proximity and size of population centers, extent of the water body's public use, and the degree of public trust all influence the level of public participation in the decision to use rotenone. An early understanding of which public groups will be involved, based on their interests, helps identify potential issues to allow development of effective responses. However, unanticipated opposition may arise (see Section 2.4.2).

2.1.4.3 Public involvement plan

A Public Involvement Plan (PIP) is critical to the success of most projects. Always develop and implement a PIP when planning an especially controversial project or when certain thresholds—likely defined by individual agencies through experience—in size, number or individuals affected, or other factors are exceeded.

The PIP should (1) identify each milestone for public involvement including dates for initial public notice, public meetings, written comments, final decisions, and notifications, (2) identify key interest individuals, groups, and agencies, (3) assign contact persons or enlist key supporters and allies in a Local Action Committee to develop and evaluate alternatives and gain support, (4) identify news media supporters and assign contact persons, (5) identify expected public response based on past reaction from the groups and individuals expected to be involved, and (6) assess methods to inform and obtain public comment, and make final decision for notification.

Make a concerted effort early in the process to identify and reasonably address controversy or criticism from the public. Critical groups might include property owners, water license holders, people using the water body for recreation, commercial interests, and animal rights advocates. Create a Local Action Committee of interested individuals to gain input and support. Encourage the public to claim "ownership" of the project plan. If a large project requires special legislative funding or is controversial, demonstration of Local Action Committee support will help legislators acquire needed funds. Contact with the public should be personal, and the project plan should indicate possible effects of the treatment on their vested interests.

Focus on the fishery problem and its unmediated impact on the public's vested interests. Do not focus on the treatment. Public involvement may be needed as early as 12–24 months before the intended treatment. Prepare a brief narrative for public distribution that summarizes the problem, proposes alternatives to correcting the problem, lists short- and long-term impacts, and delineates anticipated benefits. One or more public meetings may be necessary in the vicinity of the proposed treatment, depending on the size of the treatment and the issues and stakeholders involved. Keep a record of key actions and responses to the public involvement process.

2.1.5 Treatment plan

The preliminary treatment plan is the first snapshot of the proposed application from beginning to end. It must be completed, reviewed, and tentatively approved internally before project implementation. The preliminary plan should contain the following elements to gain an accurate assessment of necessary resources.

2.1.5.1 Physical and chemical characteristics of the water body

Prepare a general location and morphological map of the system to be treated and describe the important environmental attributes that need consideration. These may include volume or flow of water, type and density of aquatic vegetation, depth of lake, shoreline configuration and substrate, inlet and outlet flows, flushing rate, temperature, pH, dissolved oxygen, turbidity, and conductivity of water at anticipated time of treatment.

2.1.5.2 Barriers, ownership, and obstructions

Indicate and describe the barriers and obstructions to fish (hydrological) and human (topographical and legal) movement on maps. Include ownership of surrounding land and the extent and location of swampy areas and other areas that may require special treatment.

2.1.5.3 Rotenone and neutralizing agent

Describe the type of formulation, concentrations, and amounts of rotenone to be used. The rotenone concentration and formulation depend on depth, volume, water clarity, flushing rate, pH, and water temperature at the time of the proposed treatment. Typically, a lake or stream is divided into treatment zones, each with specific requirements. Specify the application rate and the amount of rotenone needed in each treatment zone. It is advisable to conduct a bioassay of the target fish to determine the concentration (efficacy) of the rotenone formulation required to accomplish to treatment goal. Assess the environmental advantages and disadvantages of natural degradation and the use of neutralizing agents. If neutralization is required, describe the type, concentration, and location of the neutralization zone. For stream treatments, distinguish between the mixing zone where fish are likely to be killed and the point where neutralization occurs and no fish are likely to be killed.

2.1.5.4 Public and commercial interests

Identify and contact public and commercial groups that use the water body, especially if it is a public or industrial water supply. It might be advantageous to involve representatives from these groups in the Local Action Committee (see Section 2.1.4.3). Document ownership of the land surrounding the water body to be treated and water licenses held, particularly for inlet and outlet streams.

2.1.5.5 Interagency responsibilities

Contact all government agencies at the local, state, provincial, or federal level that might have plans, permits, authorities, or responsibilities affected by the treatment (see Section 2.1.3.2). These include health, agriculture, parks, environment, water supply and quality, air resources or quality, and land use agencies. Include local governments (counties, cities, water reclamation districts, and conservation districts). It may be desirable to assign an agency contact person for each of these outside agencies and invite these agencies to participate on a Local Action Committee (see Section 2.1.4.3).

2.1.5.6 Logistics and preliminary schedule

Summarize the methods of operation, number of staff needed, timing, equipment needs (purchases and rentals), required permits and approvals, and biological and chemical monitoring required, and schedule each major milestone. Develop a detailed schedule for all actions required in the Intermediate Planning (see Section 2.2) and Project Implementation and Management (see Section 2.3) sections by selecting a proposed treatment date and working backward with reasonable completion dates for the milestones (see Figure 2.1). Allow for periodic assessments to amend the schedule.

2.1.5.7 Fish rescue and removal of fishing limits

The prospect of wasting fish in a treatment may prompt public concern. Consider the viability of a pretreatment salvage operation to allow the public to remove fish for their own use. Liberalization of fishing regulations can effectively address these concerns and improve public support. Ensure enough lead-time to implement regulatory changes. An alternative may be to rescue the desirable fish from the proposed treatment area for transplanting into another water body or for holding at a facility to restock once the treated water body can again support fish. These rescue alternatives are usually expensive but can be good public relations tools, especially if the public gets involved.

2.1.5.8 Restocking

Develop a restocking plan based on the proposed treatment date, management objectives, and expected rotenone degradation or neutralization time. The restocking effort should be consistent with the current FMP. The ability of federal, state, provincial, or commercial hatcheries and other populations to meet stocking needs may influence the number, size, and species of fish restocked into the treated water body.

2.1.5.9 Personnel needs

Determine personnel needs for pretreatment, treatment, and posttreatment activities based on the type and concentration of rotenone formulation and neutralizing agent prescribed and the project logistics. This estimate should include such additional assignments as preparation of the PIP, environmental analysis, fish rescue, monitoring, dead fish removal and disposal, and fish restocking.

2.1.5.10 Budget

Determine all personnel, rotenone, neutralizing agent, legal, material, and equipment (including special items) costs for the pretreatment, treatment, and posttreatment activities of the project.

2.1.5.11 Other important issues

Other important issues to consider in drafting the preliminary treatment plan may include (1) measurement of rotenone and other compounds in water, sediment, and air (see Section 3.4.3), (2) application of rotenone to public drinking or agricultural water supplies (see Section 1.3), (3) disposal of dead fish (see Section 5), (4) legal challenges, (5) threatened and endangered species (see Section 5), (6) public opposition (see Sections 2.1.4.3 and 2.4.2), (7) conflicts with other agencies (see Section 2.1.3.2), (8) applicator safety (see Section 3.3), and (9) the need for dewatering (drawing-down) a reservoir before treatment.

2.1.5.12 Contingency plans

Develop contingency plans for such possible problems as (1) incomplete dewatering of a reservoir, (2) decline in water temperature, (3) bad weather, (4) legal challenges, (5) public opposition, (6) partial kill of target species, (7) failure to neutralize rotenone, and (8) changes in stream flows (see Section 2.4.2).

2.1.5.13 Internal review

Submit the preliminary plan internally to all those who will be involved in the treatment or whose approval is needed. This will insure that adequate resources within the agency are allocated for the project, that the project is feasible and consistent with agency policy and procedures, and that tentative approval by management has been secured prior to the expenditure of planning resources.

2.2 INTERMEDIATE PLANNING

This planning stage refines the preliminary plan and clears obstacles to treatment before Project Implementation and Management (see Section 2.3).

2.2.1 Environmental analysis

2.2.1.1 Environmental quality laws

An environmental analysis (EA) may be required before a rotenone treatment depending on the applicable environmental quality laws. An EA typically focuses on environmental impacts of the project, methods of reducing environmental damage through alternatives or mitigation, and disclosure of rationale for the project. Scheduling an EA depends on the complexity of the project and the issues involved, but the document should be completed and approved before Project Implementation and Management (see Section 2.3). Ideally, the collection of information for the EA should begin sometime during the later stages of Preliminary Planning (see Section 2.1) to assist in refining the project scope. An EA may require a full year's lead time, depending on the requirements of responsible agency.

Environmental quality laws codify specific policies of federal, state, and provincial governments. These policies typically provide for (1) maintaining a quality environment, (2) identifying critical thresholds for personal health and safety, (3) encouraging systematic and concerted efforts for management of natural resources and waste disposal, (4) encouraging the enjoyment of esthetic values of natural resources, (5) preventing the elimination of fish and wildlife species, and (6) requiring government agencies to consider less environmentally damaging alternatives.

In the United States, the National Environmental Policy Act (NEPA) sets forth a systematic approach for evaluation of the environmental impacts of federal actions. Many states and provinces have similar review processes, some of which tend to place a higher value on environmental protection than on economic growth or other social considerations. For proposals subject to NEPA, an agency must evaluate and consider all reasonable alternatives and must suggest appropriate mitigation measures, but is not bound to them.

Whether subject to NEPA or state or provincial environmental quality review, the normal procedure is to conduct a preliminary analysis to determine whether the proposed treatment is categorically excluded from the need for further consideration, whether to prepare an EA which may lead to a Finding of No Significant Impact (FONSI under NEPA) or, alternatively, to prepare an environmental impact statement (EIS under NEPA). Generally, a FONSI is prepared if the project will have no significant adverse impact on the environment or if the impact can be mitigated. Alternatively, an EIS is prepared if a significant unmitigated adverse impact on the environment or an environmental change is expected. The use of federal funding for treatment on federal land will require NEPA compliance. The issues associated with the project will determine the type of NEPA compliance. Generally, a significant adverse impact is defined as a substantial adverse change in the physical or biological conditions in the affected area. The initial study should contain the following information: (1) a description of the project that includes location; (2) a checklist identification of the project's environmental effects (see Appendix C) or other similar method; (3) a discussion of suggested mitigation for significant effects; (4) a comparison of the consistency of the project with existing plans; and (5) an identification of the authors of the initial study.

2.2.1.2 Environmental studies on rotenone use in fisheries management

Several states including California, Washington, and Michigan have prepared programmatic environmental studies of rotenone use in fisheries management (WDG 1986; MDNR 1990; WDW 1992; CDFG 1994). The California Department of Fish and Game (CDFG 1994) programmatic study describes the (1) chemical properties and toxicity of rotenone and rotenone formulation constituents, (2) treatment strategies, (3) policies and criteria for rotenone use, (4) internal and external review and coordination procedures, (5) environmental assessment, (6) potential impacts, and (7) alternatives to rotenone use. In California, specific rotenone projects are supported by the programmatic document. The programmatic document serves to (1) minimize discussion duplication on issues from project to project; (2) act as a reference for the hazards of rotenone use, treatment methods, and safety procedures; and (3) generally depict the expected impacts of rotenone use. In addition to the programmatic study, site-specific environmental studies are conducted for each individual rotenone treatment in California.

2.2.1.3 Site-specific environmental analysis

The site-specific environmental study (if needed) prepared by the fish and wildlife agency proposing the project should contain the following information.

2.2.1.3.1 Fish and wildlife authorities and responsibilities—Clearly describe the need for control of fish populations and the authorities and responsibilities given the agency for this control. Include specific sections of codes or regulations discussed in Section 2.1.3.1 and important related legislation. A short history of rotenone use by the agency and others in the control of fish populations may be included.

2.2.1.3.2 Purpose of project—Clearly describe the need for an action by presenting a history of the problem, the need for correction, and the consequences of taking no action as discussed in Section 2.1.2. The purpose should be clearly related to the FMP (see Sections 2.1.1.1 and 2.1.1.2)

2.2.1.3.3 Environmental setting—Include a physical description of the area to be treated, including elevation, volume (or discharge pattern), maximum depth, and surface area. A description of the drainage basin would include tributaries, special interest areas, hydrology, and water use in the area. This section should also contain descriptions of ownership, local economics, water quality, recreation, vegetation, wildlife, and aquatic organisms as described in Sections 2.1.5.1, 2.1.5.2, and 2.1.5.4.

2.2.1.3.4 Project description—Describe the project objectives, project area and other potentially impacted areas, and proposed project plans and conditions that include reservoir dewatering. Indicate the rotenone application method, natural dissipation or neutralization (if required), and environmental monitoring. The plan for disposal of dead fish (if required), fish restocking schedule, postproject evaluation, and project cost estimate should be described according to Sections 2.1.5.7, 2.1.5.9, 2.1.5.10, and 2.1.5.11.

2.2.1.3.5 Environmental impact analysis and mitigation measures—Discuss impacts on air quality, water quality (surface and ground waters), fish, amphibians, aquatic invertebrates, birds, mammals, plants, threatened and endangered species, public health, applicators, recreation, boating, agriculture, esthetic resources, economics, and public services. Many of these issues are described in Section 5. Lessen or eliminate significant impacts by formulating mitigation measures.

2.2.1.3.6 Alternatives—The use of rotenone is one method used by an agency to meet its fisheries management objectives. Alternative methods are listed in Tables 1.1 and 1.2. Each alternative— including no treatment project—should be described along with an analysis of its potential impacts and the potential mitigation measures required to reduce impacts. Typically, the alternative analysis in an EIS under NEPA is much more detailed than in other environmental documents. An EIS must devote substantial consideration to each alternative, so reviewers may evaluate comparative merits.

2.2.2 Public and agency issue identification and notification

Concurrent with intermediate planning is the identification of issues from the public and affected agencies. The issues should become evident as the PIP (see Section 2.1.4.3) is implemented and as public meetings are conducted. Description of these issues and resolution through alternatives and mitigation will further define the project.

2.2.2.1 Agencies

Federal, state, provincial, and local agencies with jurisdictions and objectives may affect the use of rotenone. Agencies with regulatory authority (discretionary approval) over rotenone use are considered responsible agencies. In addition to the fish and wildlife agency proposing the use of rotenone, these agencies may include the U.S. Environmental Protection Agency (USEPA) and state or provincial departments of food, agriculture or pesticide regulation, public health, natural resources (i.e., forestry, lands, and parks), water quality, and environment.

The use, registration, and control of pesticides in the United States ultimately rests with USEPA. State or provincial departments that regulate pesticides, food, or agriculture enforce pesticide laws and issue licenses and certificates for pest control operations. Many states require that a licensed Agricultural Pest Control Advisor make a recommendation to use rotenone. Only a licensed, qualified applicator can supervise the application of rotenone. Agencies that regulate pesticides have had concerns with (1) safety gear, (2) safety procedures, and (3) disposal of used pesticide containers and dead fish.

State or provincial departments of health services often cooperate with the pesticide regulatory agencies in investigations of pesticide-related illnesses and develop employee safety standards for handling pesticides. Health agencies have also been delegated by the USEPA to enforce the Federal Safe Drinking Water Act through such measures as adoption of drinking water standards and monitoring regulations. Public health concerns expressed by agencies over the use of rotenone have included (1) nuisance of flies and odors created by decaying fish, (2) human consumption of dead fish containing bacteria and residues of rotenone and other compounds, (3) human consumption of drinking water containing residues of rotenone and other compounds, and (4) pesticide odors in the air.

State or provincial departments of water quality or the environment typically regulate storage and transport of hazardous wastes, disposal sites for pesticide containers, and water quality standards. Water quality and environmental agencies may establish water quality control plans that reflect water quality objectives for specific hydrologic basins.

Concerns with rotenone use from environmental agencies have included (1) impacts on beneficial uses of water, (2) maintenance of water quality standards, and (3) impacts on aquatic life other than fish (see Appendix B). Rotenone treatments may also affect the activities and interests of other agencies such as counties, cities, water reclamation districts, irrigation districts, and resource agencies.

2.2.2.2 Organizations

A number of organizations have become involved in past rotenone use projects for a variety of reasons. Environmental groups that may have an opinion on a particular project include the Sierra Club, Audubon Society, and Natural Resources Defense Council. The American Fisheries Society, angling groups like United Anglers, Trout Unlimited, and other sport fishing groups and clubs have supported projects. Conversely, groups such as People for the Ethical Treatment of Animals and Earth First have opposed the killing of fish by any means.

2.2.2.3 Public participation and notification

The PIP (see Section 2.1.4.3) should have been started in the preliminary stage and should be fully implemented in the intermediate planning stage. The extent of public participation in the planning stages of the project will be influenced by the type of project proposed and the degree of public interest and trust involved.

Release a draft EA for public review if required before agency approval is finalized (see Section 2.2.1). Notify the public of the draft EA by placing a notice in a local newspaper of general circulation in the project locale and by sending copies to interested groups identified in the PIP. Alternatively, a notice of the draft EA can be published in the state or provincial environmental bulletin, if one is regularly published. The notice should indicate the time period for public comment, a brief description and location of the project, and how to obtain a copy of the draft EA. Consider all written comments before final approval of the EA.

2.2.2.4 Press and news media

Notify local and state news media (print, radio, and television) of the proposed plan for treatment, public meetings, release of the EA, and schedule for the project. Where appropriate, the responsible agency should have a Public Affairs Officer communicate directly with the press and news media to free biologists and technicians to conduct the project (although information will originate from them). The Public Affairs Officer should act as the coordinator for all media information. Consider using Web sites to provide fast and accurate information to the public and the news media. Interviews with biologists and project personnel can produce valuable public relations benefits, but carefully coordinate them with the Public Affairs Officer. Make all agency personnel aware of the same facts and provide consistent responses when dealing the public.

2.2.2.5 Landowners and interested parties

Identify landowners in the affected area, give them information regarding the proposed project, and solicit their input. Make reasonable efforts to ensure that all interest groups are notified, and that they are on the list of interested parties.

2.2.2.6 Consistency with other management plans

Consult the management plans of water quality and environmental agencies, the U.S. Fish and Wildlife Service, U.S. Forest Service, and other affected agencies to ensure their existing management plans are considered.

2.2.3 Monitoring program

It may be desirable to monitor the application of rotenone to ensure that an effective treatment is achieved, to limit potential litigation, and to assess the impact on, and recovery of, aquatic resources. Monitoring studies can help quiet public fears about the treatment. Begin planning monitoring programs during intermediate planning after important issues, special interest areas, and the scope of the project have been clearly identified. Liquid formulations contain a variety of compounds including rotenone, dispersants, and emulsifiers (see Section 5.3), and their dissipation in the environment over time may be of interest. The number and location of sample sites and sampling frequencies will vary with each treatment. Several monitoring methods have been used for rotenone treatments, including analyzing water, sediment, and air samples for residues of rotenone and other compounds and assessing the impact of the treatment on biological (e.g., fish, amphibian, and invertebrate) resources. Procedures for monitoring are described in Section 3.4. An overview and objectives of monitoring are given below.

2.2.3.1 Water monitoring

Collect water samples to document the initial concentrations and degradation over time of rotenone and associated compounds (i.e., rotenolone, piperonyl butoxide, volatile organic compounds [trichloroethane, trimethylbenzenes, toluene, xylenes, and ethyl benzene], and semivolatile organic compounds [naphthalene and methyl naphthalene]) in surface water and groundwater. Pretreatment samples should be collected, followed by collection of posttreatment samples at specific time intervals. Rotenone can persist from several days to several months depending on water temperature and initial concentration. Information from the monitoring process will confirm that lethal levels of rotenone were applied in the target area and that rotenone has degraded before restocking with fish. There are water quality objectives and drinking water standards for many of the compounds in rotenone formulations (see Section 5). Samples from neutralization zones can confirm degradation of rotenone over time and distance. Groundwater monitoring studies (see Section 5) have never detected contamination from a rotenone treatment.

2.2.3.2 Air sampling

Air samples may be collected to document initial concentrations and dissipation of rotenone, rotenolone, and volatile and semivolatile organic compounds over time. Air sampling is particularly helpful to document movement of volatile organic compounds from the treatment area. Permissible airborne exposure limits (PELs) have been established for worker protection for rotenone and several volatile and semivolatile organic compounds (see Section 5). This information is useful in verifying the safety of the treatment.

2.2.3.3 Sediment sampling

Rotenone, rotenolone, and the semivolatile organic compounds (naphthalene and methyl naphthalene) are transient in sediment from treatment areas. The reduction of concentrations of these materials in sediments appears to lag about 1 to 2 weeks behind the reduction of concentrations in water (see Section 5). Rotenone is not expected to be biologically active in the sediment or upon resuspension of the sediment; however, it may be prudent to wait until residues in the sediment degrade to nondetectable levels before restocking fish.

2.2.3.4 Biological monitoring

Live-cages containing sentinel fish (normally the target species or a species with the same or less sensitivity to rotenone than target species) may be placed vertically and horizontally throughout the treatment area to serve as indicators that lethal concentrations of rotenone were applied. Placing sentinel fish in adjacent water as a control will verify that mortality is due to rotenone and not other factors. Caged fish placed downstream of neutralization areas will confirm successful dissipation of rotenone. Use this technique before restocking to assure survival of fish in the treatment area.

2.3 Project Implementation and Management

The last stage before treatment is to finalize plans for all operations associated with the project. Develop a crisis management plan for large, high profile, or controversial projects well before the rotenone treatment is implemented (see Section 2.4.2). A project schedule and structure are needed to organize large projects. For large, complex, or controversial projects, you may wish to employ a version of an Incident Command System (ICS) to organize the various functions (NIFC 1994; see Appendix D). For some treatments no ICS may be needed, while others may require many of the ICS's functional elements. Regardless of the structure chosen, only qualified personnel (e.g., Pest Control Advisor or Qualified Applicator) knowledgeable of rotenone and these guidelines must supervise the treatment. Administrative approval for the treatment is obtained from the highest possible level in the fish and wildlife agency commensurate with the scope of the project, ensuring adequate agency review. Assignments should be made for completing the project-specific work plans by specific dates. Each plan must contain sufficient information and detail so others can use the plan and complete the needed activity without assistance. Complete all plans at least one month before treatment to allow for sufficient review and approval time. Also, ensure that all approvals from other agencies have been obtained and documented (see Section 2.1.3). Depending on the size, complexity, and location of the treatment, plans may be needed for some or all of the following operations.

2.3.1 Public involvement plan

Update and modify the PIP (see Section 2.1.4.3) as necessary based on public input received to date.

2.3.2 Fish rescue and removal of fishing limits plan

Outline the actions and schedule of events needed for a fish rescue and possible removal of fishing limits (see Section 2.1.5.7) if these are elements of the project.

2.3.3 Rotenone application plan

Outline the actions and schedule of events required to treat the water body (see Section 2.1.5.3). This may include the developing a schedule for dewatering an impoundment before treatment.

2.3.4 Monitoring plan

Outline the actions and schedule of events required to monitor the treatment (see Section 3.4; see Appendix E for example).

2.3.5 Site safety plan

Indicate the safety gear and practices required to ensure a safe and contamination-free application. An on-site safety training exercise before application (see Section 3.3; Appendix F for example) may be useful. Include a safety contingency plan for accidents that involve personnel (e.g., determine nearest hospital, type of first aid required, etc.).

2.3.6 Site security plan

Indicate where the operation will be staged and the security required for the storage of equipment, rotenone, and other materials, and for the safety of personnel. Indicate the actions and schedule of events required to attain the level of security required. Consider controlling public access (angling and boating) to the treatment area to limit liability and interference.

2.3.7 Fish removal and disposal plan

Indicate if dead fish will be removed. If they are to be removed, indicate by whom, when, where, and how the fish will be disposed of and if approvals or permits are needed. Indicate the actions and schedule of events required to remove and dispose of fish.

2.3.8 Spill contingency plan

Indicate how the rotenone and the detoxifier of rotenone, potassium permanganate, will be transported to the site, how it will be stored on-site until application, what precautions will be taken to prevent spills, and how spills will be neutralized and cleaned up if the need arises. Also indicate the names and numbers of government agencies to be notified in the event of a spill (see Appendix G for example).

2.3.9 Site neutralization plan

Indicate the location of the neutralization site, anticipated dates of operation, type of neutralizing agent, desired concentration, estimated amount of neutralizing agent required, application device, frequency of monitoring, and personnel requirements (see Section 2.1.5.3).

2.3.10 Communication plan

Detail how personnel will communicate with one another and the news media during the treatment. This is especially important when coordinating rotenone application at booster stations and neutralization stations at various locations along a stream.

2.3.11 Restocking plan

Indicate the actions and schedule of events required to restock the water body (see Section 2.1.5.8)

2.3.12 Records maintenance plan

Written records of all decisions and significant events that involve planning and execution of the project must be kept for a posttreatment debriefing and possible litigation.

2.4 TREATMENT

The planning process outlined in Sections 2.1 through 2.3 should have prepared the fish and wildlife agency technically, politically, socially, and legally for the rotenone treatment which is described in Section 3. The agency must understand the techniques for rotenone use in Section 3 to provide a sound foundation for planning a treatment. It is imperative that these planning activities occur before treatment, commensurate with guidance in Section 2.0. The treatment occurs at the end of the planning process. Once the treatment has begun, the success of the project-specific work plans in meeting objectives should be monitored and, if necessary, amended to achieve the necessary objectives.

2.4.1 Monitor ongoing public relations

The PIP drafted during preliminary planning should be fully implemented before treatment. Monitor ongoing public relations during and immediately following the treatment.

2.4.2 Crisis management strategies

Crises typically result from adverse public reaction to, or excessive media interest in, an unplanned event during or following the treatment. Examples from recent rotenone treatments include (1) failure of the neutralization operation, killing fish outside of the treatment area, (2) strong "pesticide" odors resulting in complaints and reported illnesses, (3) chemicals in the rotenone formulation persisting for longer periods than anticipated, and (4) failure of the treatment to accomplish the stated objectives.

2.4.2.1 Crisis management plan

This action plan prepares you for any negative development that may jeopardize the rotenone application or its favorable outcome. Before treatment, identify a crisis team that will act as an early alert group to develop the situation response and then use the appropriate crisis team participants and support groups.

- The crisis team
  The crisis team should include (1) early alert members (i.e., persons who can handle the crisis and devote exclusive time to the crisis), (2) primary response members (e.g., technical experts in various disciplines), and (3) secondary response members (e.g., high-level persons in your agency, elected officials).
- Situation response
  The situation response comprises the following steps: (1) define the problem and scope; (2) identify targets and issues; (3) select appropriate crisis team; (4) gather facts; and (5) identify a spokesperson.
- Targets and issues
  Consider public relations and natural resources (all aspects of involvement—environment, birds, animal rights and welfare, etc.) when identifying the targets and issues.
- Support groups
  Support groups normally consist of members of (1) research groups, (2) sports clubs, associations, and organizations, and (3) advocate and regulatory governmental agencies. Gain support before the incident becomes a major crisis.

2.4.2.2 Managing a crisis

The following is a step-by-step method of managing a crisis.

- Define the real problem
  Gauge public actions and opinions, perhaps by using a newspaper clipping service. Focus on long-term consequences; do not focus on the details. Delegate details to support groups.
- Identify a crisis team
  Choose the team carefully for the situation from your preselected lists (early alert, primary response, and secondary response members). These individuals should devote themselves entirely to the crisis. Do not delay; act immediately!
- Resist combative instincts
  No matter what circumstances produced the crisis, keep control, or control of the situation will be lost.
- Centralize control of information
  Centralize control of information that is released to the public and keep the message consistent and clear.
- Communicate and negotiate at the highest level of authority
  Communicate and negotiate at the highest level of authority possible. Follow the chain of command and brief all involved. Keep administration informed.
- Contain problem quickly
  Contain the problem quickly and stop the erosion of public confidence.

2.4.2.3 Media relationships

- Be prepared
  Success with the media depends on preparation. Organize facts, anticipate questions, and plan concise answers. Write down two or three important messages (specific information you want the public to hear) in advance.
- Be honest
  Provide accurate information and be completely honest. Do not try to fool reporters or the public. If an answer is not known, admit it, but make the effort to provide the answer later. Do not speculate!
- Understand the media
  Make sure the spokesperson understands the needs of the different media. Be sensitive to reporters' deadlines. They have a job to do.
- Be accessible and return calls quickly
- Stick to the facts
  Stay with the topic at hand. Keep the interview on track by emphasizing essential points.
- Be brief
  Answer in quotable statements of a duration of 20 seconds or less.

- Educate
  Explain scientific and technical information carefully. Use plain English, not jargon.
- Avoid confrontation
  Stay calm! Do not argue, lose composure, or confront reporters. If questions are stated inaccurately or combatively, simply correct the remarks in the answer and do not repeat inflammatory words used in the question.
- Demonstrate leadership
  Let the media know the situation is under control and emphasize what is being done to correct any problem. Offer positive actions and solutions.
- Interview composure
  Take charge, anticipate questions, develop a key message in advance, stick to the facts, and remain calm.

Appendix B

Memorandum of Understanding between the California Department of Fish and Game and the California Regional Water Quality Control Board—Lahontan Region on Rotenone Use in Fisheries Management

USE OF THE FISH TOXICANT ROTENONE

Memorandum of Understanding
Between the
California Water Quality Control Board
Lahontan Region
and the
California Department of Fish and Game

This Memorandum of Understanding (MOU) is entered into by and between the California Regional Water Quality Control Board, Lahontan Region (hereinafter "Regional Board") and the California Department of Fish and Game (hereinafter "Department"). This MOU constitutes an agreement regarding the Department's use of the fish toxicant rotenone.

The Regional Board has been invested by the people of the State with the responsibility to maintain water quality in the Lahontan Region. The Regional Board performs this function by developing and enforcing policies and standards for water quality in the Region. These standards and policies are contained in the Water Quality Control Plans for the Lahontan Region (Basin Plans), and apply to any activity which might affect water quality.

The Department of Fish and Game has been invested by the people of the State with the responsibility to carry out a variety of fishery management activities. These activities are designed to protect and maintain valuable sport fisheries and aquatic ecosystems. The Department is also responsible under State and Federal law for the restoration and protection of threatened and endangered species.

The Department has determined that in order to carry out effective management programs, it is sometimes necessary to completely eliminate existing fish populations in designated areas. This practice provides optimum conditions for propagation of healthy, desirable fish. One method of eliminating existing fish populations has been through the application of rotenone. The Department has assessed the practicality and potential effectiveness of alternatives to rotenone use, and concluded that in certain situations the use of rotenone will be critical for success in achieving its management objectives. The Department has therefore proposed the use of rotenone for management purposes. However, the use of rotenone entails a number of short-term impacts on the quality of treated waters, and is therefore of concern to the Regional Board.

The Regional Board amended its Basin Plans in 1990 to include a policy regarding the Department's use of rotenone. The policy applies to rotenone use for the following types of activity: 1) the restoration or enhancement of threatened or endangered species; 2) the control of fish diseases; and 3) the eradication of harmful prohibited or harmful "exotic" (i.e., introduced) species. The policy may also be extended to apply to other types of fishery management projects on a case-by-case basis when there is sufficient justification.

The two agencies agreed to enter into this Memorandum of Understanding in order to better meet their mutual objectives. This Memorandum was subsequently prepared, and applies only to proposed rotenone projects that fall within the provisions of the Basin Plans.

IT IS HEREBY AGREED THAT:

I. For each proposed rotenone treatment project, the Department will provide the following specified items and project specific information to the Regional Board in advance of the proposed project start-date. In providing the specified items the submittal may reference other pertinent documents when those documents have been previously submitted to the Regional Board:

A. Project Map
Clearly delineating the proposed treatment area, approximate locations of any expected rotenone drip stations and project boundaries, and approximate location of the detoxification station.

B. Project Officials
Names of Department contact person for the project, and the project leader.

C. Site selection

1. Completed DFG form 740[1] for the project (Attachment A hereto) includes criteria used to judge project priority, and general information about project site.
2. Completed DFG form 680[1] for the project (Attachment B hereto). Contains information regarding spill contingency plan, information plan, potable water supplies, monitoring programs, treatment rates, and government and public notification.
3. List of steps taken to fulfill the Department's internal review procedure (as outlined in section VI of the EIR), including a list of officials who reviewed and approved the documents, and dates of review and approval.
4. A list of specific target organisms.
5. Copies of all California Environmental Quality Act (CEQA) documents and accompanying information, and certification of CEQA compliance.

D. Public and Governmental Notification

1. Dates of any public hearings.
2. List of potentially affected sources of potable surface and groundwater intakes, and description of steps that were taken to identify these intakes.
3. Plans for notification of potentially affected residents or other potential users of the affected area.

E. Project Planning

1. The expected project start-date.
2. The Department's project-specific assessment of the potential effectiveness of using alternatives to rotenone, according to detailed evaluation procedure described in the Department's 1985 Programmatic Environmental Impact Report *Rotenone Use for Fisheries Management*, Section IV (hereinafter referred to as EIRII).
3. The name and manufacturer of the commercial rotenone formulation to be used.
4. Lot numbers of formulation that will be used.
5. Results of organic analytical scans for each lot of formulation that will be used. The scans must be performed by a laboratory certified for hazardous waste analysis.

[1] Forms DFG 680 and 740 available from California Department of Fish and Game, 1701 Nimbus Road, Suite F, Rancho Cordova, California 95670, USA.

6. Planned method of application (for example, by helicopter, drip station, or by hand-spraying).
7. Planned application rate and the rationale for choosing this rate considering the sensitivities of target organisms.
8. Methods and calculations (should consider site volumes, surface areas, and flows), and plans for calibrating application equipment to achieve planned application rate.
9. Methods and calculations (should consider flows and other applicable site conditions), and plans for calibrating detoxification equipment to achieve minimum effective concentration of potassium permanagate (a detoxifying agent).
10. Specific quantities of chemicals to be transported, intended methods of transportation and storage.
11. Spill contingency plan (to include containment and cleanup measures, emergency notification plans and emergency chain of command).
12. Plans for disposal of dead fish.
13. List of criteria for determining the need to begin detoxification, and for determining that detoxification can be safely discontinued.
14. Details of how the Department intends to provide drinking water to residents, it monitoring indicates that water supplies have been contaminated (this item is required as part of the submittal only if there is a significant risk of contamination to drinking water supplies).
15. Detailed treatment plan if the treatment is to take place over two or more days. This plan should describe which portions of the project site will be treated in what order and in what time frame.

F. Monitoring

1. Scaled map and description of proposed ground and surface water sampling stations and sampling schedules.
2. Copy of intended reporting format (tables) for reporting monitoring results (including format for reporting results of Quality Control measures such as replicates, spikes, and reference standards).
3. Name, address, phone number and contact person of the laboratory that will be conducting chemical analysis of monitoring samples.
4. Proof of laboratory certification.
5. Copy of the laboratory Quality Assurance/Quality Control program.

II. The Department will work with the U.S. Fish and Wildlife Service in developing more effective rotenone formulations with less objectionable 'inert' ingredients. The Department will provide annual updates to the Regional Board on these developments.

III. The Department has estimated that fifteen to thirty minutes of contact time is required between rotenone-treated water and potassium permanganate for complete detoxification. Therefore the Department will conduct a travel time study before each project to estimate the distance covered in a 30-minute travel time starting at the detoxification station. The results of the study will be submitted to the Regional Board Executive Officer at least seven (7) calendar days prior to treatment. Project boundaries will then be defined as encompassing the treatment area, the detoxification area, and the area downstream of the detoxification station up to the thirty-minute-travel-time point.

IV. Within two years of the last treatment date for a given project, the Department will send a qualified biologist to the project site to assess the condition of the treated waters and the condition of fish and invertebrate populations in those waters, and certify that beneficial uses have been restored.

V. Deviations from project-specific plans may occur only upon prior mutual agreement between the two parties.

VI. For the purposes of this memorandum, proposed projects will be defined either as "new projects", "repeat projects", or "emergency projects." Because rotenone projects typically require treatment during several consecutive years, projects that have not yet undergone their first year of treatment will be considered "new projects." Those projects that have undergone treatment in previous years, and that have been reviewed in conformance with the procedures outlined in this memorandum, will be considered "repeat projects." (The current Upper Truckee River/ Meiss Meadows project will initially be considered a "repeat project.") Projects requiring immediate action, and which have been officially designated as emergencies by the Director of the Department, will be considered "emergency projects." All currently planned projects will be considered new projects for the purposes of this memorandum, except for the Upper Truckee River/Meiss Meadows project, which will be considered a repeat project.

VII. The Department will provide the items and information described in Section I of this memorandum no later than sixty (60) calendar days before the anticipated start of any now project, thirty (30) calendar days before the start of any repeat project, and seven (7) calendar days before the start of any emergency project. The Department will notify the Regional Board immediately upon any Departmental decision to initiate an emergency rotenone project.

VIII. Based on the information submitted under Section I above, the Regional Board Executive officer will determine whether the proposed project is consistent with applicable provisions of the Basin Plans and meets the following criteria:

1. The Basin Plan limitations for chemical residue levels resulting from rotenone use can be met.
2. The planned treatment protocol will result in the minimum discharge of chemical substances that ran reasonably be expected for an effective treatment.
3. Chemical transport, spill contingency plans, and application methods will adequately provide for protection of water quality.
4. Suitable measures will be taken to notify the public, and potentially affected residents.
5. Suitable measures will be taken to identify potentially affected sources of potable surface and groundwater intakes, and to provide potable drinking water if necessary.
6. A suitable monitoring program will be followed to assess the effects of treatment on surface and groundwater, and on bottom sediments.
7. The Department has gone through the process required by the California Environmental Quality Act (CEQA).
8. The Department has gone through the processes set forth in its Environmental Impact Report "Rotenone Use for Fisheries Management" (1985).

16. The chemical composition of the rotenone formulation has not changed significantly in such a way that potential hazards may be presented which have not been addressed.
17. Plans for disposal of dead fish are adequate to protect water quality.

The Executive Officer may request revisions to the proposed plans, or additional information from DFG, before making the determination. The Executive Officer must either notify the Department of his determination, request specified additional information, or request revisions to the plans, within the following time schedule:

i) new projects—within thirty (30) calendar days from receipt of the submittal.
ii) repeat projects—within fifteen (15) calendar days from receipt of the submittal.
iii) emergency projects—within four (4) calendar days from receipt of the submittal.

If additional information or revisions are requested the Department will have seven (7) calendar days for new or repeat projects, or two (2) calendar days for emergency projects, from the date that the request is made, to provide the information to the Executive Officer. If the Department fails to meet this deadline, the Executive Officer may, at his/her discretion determine that the proposed project has failed by default to meet the criteria.

If additional information or revisions are provided according to schedule, the Executive Officer will have seven (7) calendar days for new or repeat projects, or two (2) calendar days for emergency projects, from the date that the requested material is received, to make the final determination.

If the Executive Officer fails to make the determination within the allotted time, the determination will be made by default, and the project will be considered to have successfully met the criteria.

IX. Should the Executive Officer determine that a proposed project cannot meet applicable provisions of the Basin Plan, or cannot meet the criteria listed in Section VIII above, he/she will notify the Department immediately. The two parties may then elect to schedule a meeting to settle any disputed issues. The meeting shall take place within seven (7) calendar days of the determination for new or repeat projects, or within two (2) calendar days of the determination for emergency projects. The meeting will take place between the Executive Officer of the Regional Board and the Director of the Department, or their designated representatives. If the dispute is settled through this meeting, the Executive Officer will have two (2) calendar days to make a new determination. However, if any issues remain unresolved after such a meeting, or if the parties should decline to participate in such a meeting, either party may choose at that time to abrogate this memorandum in writing. In such a case, this entire agreement will be immediately nullified and each party will then take the course of action, which it considers appropriate. If nullified, this memorandum may be renewed at a later time by the written consent of both parties. If the Department wishes, it may petition the Regional Board within seven (7) calendar days of receiving the determination, to review the determination of the Executive Officer.

X. To facilitate evaluation of monitoring results, monitoring reports provided to the Regional Board by the Department will include the specific time of day at which each sample was collected. A separate report will be included which notes the date on which each sample was analyzed, and which provides the results of laboratory Quality Control checks that were performed on that day (such as checks for accuracy, precision, and percent recovery).

XI. The Department will provide all monitoring results, and results of Quality Control checks, to the Regional Board within sixty (60) calendar days of the last day on which samples are taken for the project. If the Regional Board takes any samples for the project, they will provide the analytical results from those samples to the Department, within sixty (60) calendar days of the last sampling date.

XII. This Memorandum of Understanding shall become effective after execution and shall remain in force until terminated by written notice by either party.

XIII. This Memorandum of Understanding may be ammended as mutually agreed upon by the Department of Fish and Game and the Regional Board.

XIV. This memorandum does not constitute a waiver, on the part of either party, of any existing rights or statutory obligations. This agreement is designed to operate as a complement to each party's legislative mandates and policies, without infringing the right of either party to fulfill its legal responsibilities.

XV. All notices and communications under this Memorandum of Understanding shall be addressed to the following:

Peter Bontadelli
Director
1916 Ninth St.
Sacramento, CA 95814

Harold J. Singer
Executive Officer
P.O. Box 9428
South Lake Tahoe, CA 95731

This Memorandum of Understanding is executed on the date of the most recent signature below, by the following authorized representatives of the parties.

Peter Bontadelli
Director
California Dept. of Fish & Game

Date

Harold J. Singer
Executive Officer
California Regional Water Quality Control Bd.
Lahontan Region

Date

APPENDIX C

ENVIRONMENTAL CHECK LIST

ENVIRONMENTAL CHECKLIST FORM

PROJECT LOCATION: ____

CITY COUNTY

PROJECT ADDRESS: ____

DESCRIPTION OF PROJECT: ____

ENVIRONMENTAL IMPACTS: ____

(Form requires that an explanation of all "yes" and "maybe" answers be provided along with this checklist, including a discussion of ways to mitigate the significant effects identified. You may attach separate sheets with the explanations on them.)

	Yes	Maybe	No
I. EARTH. Will the proposal result in:			
a) Unstable earth conditions or in changes in geologic substructures?	☐	☐	☐
b) Disruptions, displacements, compaction, or overcovering of the soil?	☐	☐	☐
c) Change in topography or ground surface relief features?	☐	☐	☐
d) The destruction, covering, or modification of any unique geological physical features?	☐	☐	☐
e) Any increase in wind or water erosion of soils, either on or off the site?	☐	☐	☐
f) Changes in deposition or erosion of beach sands, or changes in siltation, deposition, or erosion which may modify the channel of a river or stream or the bed of the ocean or any bay, inlet, or lake?	☐	☐	☐
g) Exposure of people or property to geologic hazards, such as earthquakes, landslides, mudslides, ground failure, or similar hazards?	☐	☐	☐
II. AIR. Will the proposal result in:			
a) Substantial air emissions or deterioration of ambient air quality?	☐	☐	☐
b) The creation of objectionable odors?	☐	☐	☐
c) Alteration of air movement, moisture, or temperature, or any change in climate, either locally or regionally?	☐	☐	☐
III. WATER. Will the proposal result in:			
a) Changes in currents, or the course of direction of water movements, in either marine or freshwaters?	☐	☐	☐
b) Changes in absorption rates, drainage patterns, or the rate and amount of surface runoff?	☐	☐	☐
c) Alterations to the course or flow of floodwaters?	☐	☐	☐
d) Changes in the amount of surface water in any water body?	☐	☐	☐
e) Discharge into surface waters or in any alteration of surface water quality, including, but not limited to, temperature, dissolved oxygen, or turbidity?	☐	☐	☐
f) Alteration of the direction or rate of flow of groundwaters?	☐	☐	☐
g) Change in the quantity of groundwaters, either through direct additions or withdrawals, or through interception of an aquifer by cuts or excavations?	☐	☐	☐

	Yes	Maybe	No
h) Substantial reduction in the amount of water otherwise available for public water supplies?	☐	☐	☐
i) Exposure of people or property to water-related hazards such as flooding or tidal waves?	☐	☐	☐
IV. PLANT LIFE. Will the proposal result in:			
a) Change in the diversity of species or number of any species of plants (including trees, shrubs, grass, crops, and aquatic plants)?	☐	☐	☐
b) Reduction of the numbers of any unique, rare, or endangered species of plants?	☐	☐	☐
c) Introduction of new species of plants into an area resulting in a barrier to the normal replenishment of existing species?	☐	☐	☐
d) Reduction in acreage of any agricultural crop?	☐	☐	☐
V. ANIMAL LIFE. Will the proposal result in:			
a) Change in the diversity of species or numbers of any species of animals (birds; land animals, including reptiles; fish and shellfish; benthic organisms; or insects)?	☐	☐	☐
b) Reduction of the numbers of any unique, rare, or endangered species of animals?	☐	☐	☐
c) Introduction of new species of animals into an area resulting in a barrier to the migration or movement of animals?	☐	☐	☐
d) Deterioration to existing fish or wildlife habitat?	☐	☐	☐
VI. NOISE. Will the proposal result in:			
a) Increases in existing noise levels?	☐	☐	☐
b) Exposure of people to severe noise levels?	☐	☐	☐
VII. LIGHT AND GLARE. Will the proposal:			
a) Produce new light or glare?	☐	☐	☐
VIII. LAND USE. Will the proposal result in:			
a) Substantial alteration of the present or planned land use of an area?	☐	☐	☐
IX. NATURAL RESOURCES. Will the proposal result in:			
a) Increase in the rate of use of any natural resources?	☐	☐	☐
X. RISK OF UPSET. Will the proposal involve:			
a) A risk of an explosion or the release of hazardous substances (including, but not limited to: oil, pesticides, chemicals, or radiation) in the event of an accident of upset condition?	☐	☐	☐
b) Possible interference with an emergency response plan or an emergency evacuation plan?	☐	☐	☐
XI. POPULATION. Will the proposal:			
a) Alter the location, distribution, density, or growth rate of the human population of an area?	☐	☐	☐
XII. HOUSING. Will the proposal:			
a) Affect existing housing or create a demand for additional housing?	☐	☐	☐
XIII. TRANSPORTATION/CIRCULATION. Will the proposal result in:			
a) Generation of substantial additional vehicular movement?	☐	☐	☐
b) Effects on existing parking facilities or demand for new parking?	☐	☐	☐

	Yes	Maybe	No
c) Substantial impact upon existing transportation systems?	◻	◻	◻
d) Alterations to present patterns of circulation or movement of people or goods?	◻	◻	◻
e) Alterations to waterborne, rail, or air traffic?	◻	◻	◻
f) Increase in traffic hazards to motor vehicles, bicyclists, or pedestrians?	◻	◻	◻
XIV. PUBLIC SERVICES. Will the proposal have an effect upon, or result in a need for new or altered governmental services in any of the following areas:			
a) Fire protection?	◻	◻	◻
b) Police protection?	◻	◻	◻
c) Schools?	◻	◻	◻
d) Parks or other recreational facilities?	◻	◻	◻
e) Maintenance of public facilities, including roads?	◻	◻	◻
f) Other governmental services?	◻	◻	◻
XV. ENERGY. Will the proposal result in:			
a) Use of substantial amounts of fuel or energy?	◻	◻	◻
b) Substantial increase in demand upon existing sources of energy or require the development of new sources of energy?	◻	◻	◻
XVI. UTILITIES and SERVICE SYSTEMS. Will the proposal result in a need for new systems or substantial alterations to the following utilities:			
a) Power or natural gas?	◻	◻	◻
b) Communications systems?	◻	◻	◻
c) Water?	◻	◻	◻
d) Sewer or septic tanks?	◻	◻	◻
e) Storm water drainage?	◻	◻	◻
f) Solid waste and disposal?	◻	◻	◻
XVII. HUMAN HEALTH. Will the proposal result in:			
a) Creation of any health hazard or potential health hazard (excluding mental health)?	◻	◻	◻
b) Exposure of people to potential health hazards?	◻	◻	◻
XVIII. AESTHETICS. Will the proposal result in:			
a) The obstruction of any scenic vista or view open to the public?	◻	◻	◻
b) The creation of an aesthetically offensive site open to public view?	◻	◻	◻
XIX. RECREATION. Will the proposal result in:			
a) Impact upon the quality or quantity of existing recreational opportunities?	◻	◻	◻
XX. CULTURAL RESOURCES. Will the proposal:			
a) Result in the alteration of or the destruction of a prehistoric or historic archaeological site?	◻	◻	◻
b) Result in adverse physical or aesthetic effects to a prehistoric or historic building, structure, or object?	◻	◻	◻
c) Have the potential to cause a physical change which would affect unique ethnic cultural values?	◻	◻	◻
d) Restrict existing religious or sacred uses within the potential impact area?	◻	◻	◻

XXI. MANDATORY FINDINGS OF SIGNIFICANCE.

a) Potential to degrade: Does the project have the potential to degrade the quality of the environment, substantially reduce the habitat of a fish or wildlife species, cause a fish or wildlife population to drop below self-sustaining levels, threaten to eliminate a plant or animal community, reduce the number or restrict the range of a rare or endangered plant or animal, or eliminate important examples of the major periods of history or prehistory?

b) Short-term: Does the project have the potential to achieve short-term, to the disadvantage of long-term, environmental goals? (A short-term impact on the environment is one which occurs in a relatively brief, definitive period of time. Long-term impacts will endure well into the future.)

c) Cumulative: Does the project have impacts which are individually limited but cumulatively considerable? (A project may impact on two or more separate resources where the impact on each resource is relatively small but where the effect on the total of those impacts on the environment is significant.)

d) Substantial adverse: Does the project have environmental effects which will cause substantial adverse effects on human beings, either directly or indirectly?

XXII. DISCUSSION OF ENVIRONMENTAL EVALUATION: (This section may be filled out by using narrative or by using a form.)

XXIII. DISCUSSION OF LAND USE IMPACTS. (An examination of whether the project would be consistent with existing zoning, plans, and other applicable land use controls.)

Appendix D

Incident Command System Organization

INCIDENT COMMAND SYSTEM ORGANIZATION

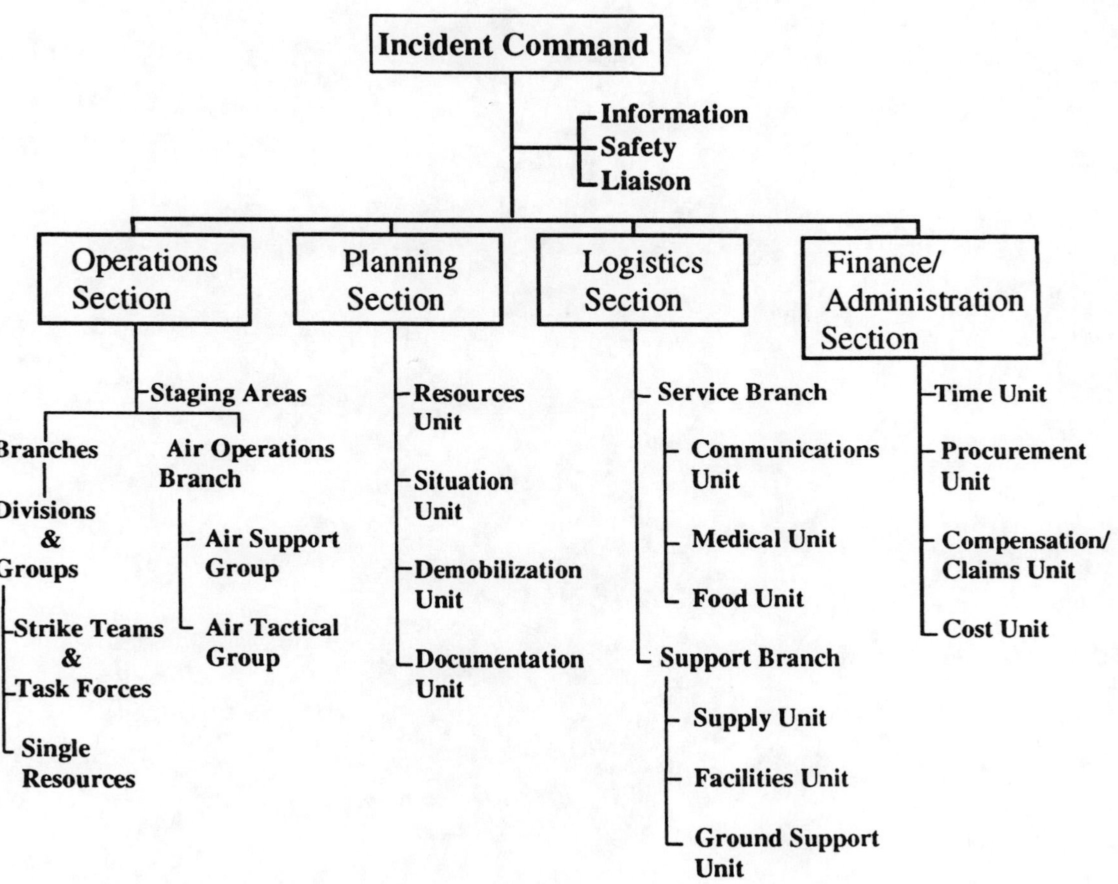

Appendix E

Environmental Monitoring Plan

Lake Davis (Plumas County) Northern Pike Eradication Project Draft Environmental Recovery Monitoring Plan California Department of Fish and Game

This monitoring plan consists of four elements: (1) a chemical monitoring plan; (2) a toxicological monitoring plan; (3) an invertebrate monitoring plan; and (4) a taste and odor monitoring plan.

I. Chemical Monitoring

A. Objectives

1. Confirmation that the rotenone concentrations present in the lake were sufficient to eliminate the target species;
2. Confirmation that complete degradation of rotenone and other materials in Nusyn-Noxfish® has occurred prior to the restocking of fish and resumption of public contact with the lake waters;
3. Confirmation that toxic concentrations of rotenone do not impact Big Grizzly Creek downstream of Big Grizzly Dam;
4. Confirmation that contamination of wells in the area has not occurred;
5. Confirmation that there are no detectable residues of rotenone or other materials in Nusyn-Noxfish® in Lake Davis water before it is processed by the Plumas County Flood Control District Water Treatment Plant as potable water;
6. Confirmation that any water quality impairment caused by the treatment (e.g., biological oxygen demand) is mitigated to not negatively impact drinking water quality or survival of restocked fish.

B. Materials and Methods

Water samples will be collected using the methods of Harrington and Finlayson (1988) for surface water and sediment and Sava (1986) for groundwater. Analysis for rotenone and rotenolone concentrations will be performed by a California Department of Fish and Game (CDFG) laboratory using methods described in Dawson et al. (1983). Analysis for the other organic constituents and potential contaminants, of the formulation will be performed by a CDFG laboratory or a private laboratory using U.S. Environmental Protection Agency-approved methods. Replicate samples will be taken for analysis by Department of Health Services (DHS). The numbers of samples given in Tables A–G does not include samples taken for the DHS.

Preproject monitoring will occur in June, July, August, and September/October 1997. Post-project monitoring will occur as described below with the first day of treatment being Day 0.

1. Sample Collection and Storage
   a. Water samples
      i) Water sampling for rotenone and rotenolone will utilize 500-ml amber glass bottles filled to capacity and sealed with teflon-line caps. Water samples for the non-rotenoid organic constituents of the formulated rotenone product will be collected in 1000 ml amber glass bottles and volatile organic glass vials. Grab surface water samples will be taken from the lake, creek, and water treatment plant. Subsurface water samples will be collected using a Kemmerer bottle from mid-depth and one foot above the bottom in Lake Davis.

ii) Water temperatures and the dissolved oxygen content of surface waters will be recorded at the time of sample collection using a YSI® model 57 oxygen meter. Water samples for pH, hardness, total organic carbon, ammonia (or total nitrogen), conductivity, and alkalinity determinations will be collected in 1-L high-density polyethylene bottles from surface water, water treatment plant, and groundwater.

iii) Samples taken for Biochemical Oxygen Demand (BOD) will be taken in 1-L high density polyethylene jars and transported immediately to the Water Pollution Control Laboratory (WPCL) for analysis.

iv) Bacteriological analysis will be performed on groundwater samples and samples taken in the treatment plant. Sampling protocol specified by the DHS will be followed.

b. Sediment samples

i) Sediment samples will be taken using a sediment core sampler with plastic liner sleeves. The top 6 in of hydrosoil will be analyzed.

c. Storage

All samples will be stored on ice at a temperature of 4°C and transported to the Pesticide Investigations Unit (PIU). Sediment samples will be frozen pending analysis. The samples will be transferred to the WPCL in Rancho Cordova for analysis. The samples for other organic compounds may be transferred to a contract laboratory for analysis. Field blanks will be utilized.

2. Analyses

a. Rotenone and Rotenolone—A 500-ml aliquot of a water sample buffered to pH 5 will be filtered through a preconditioned Sep Pak® at a rate not to exceed 40 ml/min using a vacuum pump according to the method of Dawson et al. (1983). Rotenone and rotenolone will be extracted from the Sep Pak® with methanol and analyzed on a Varian® model 500 high-performance liquid chromatograph on an MCH 10 reverse-phase column with methanol:water (75:25) mobile phase and wavelength of 275 nm. The MDL values for rotenone and rotenolone are 2 g/L for a 500-ml sample volume. Coefficient of variations for duplicate samples from CDFG studies conducted in 1986 and 1987 using this method ranged from 3 to 20%. Rotenone was relatively stable for six days when stored at 4°C in the absence of light (Harrington and Finlayson 1988). The objective is to have all samples extracted in methanol within 48 h after collection.

b. Other organic compounds—These samples will be analyzed by a State of California certified laboratory using EPA methods 502.2 and 8310 for water and EPA methods 8270B, and 8260A for sediment.

c. BOD—These samples will be analyzed by a State of California certified laboratory using EPA method 5210B.

d. Bacteriological—These samples will be analyzed for total and fecal coliform bacteria by a State of California certified laboratory using methods specified by the DHS.

e. Water Quality—These samples will be analyzed using standard APHA methods.

3. Sample Security and Data Handling

Each sample collected will be accompanied by a Chain of Custody form documenting the sequence of transfer from sample generation to chemical analyses.

The form will include location codes, sampling dates and times, sample description, and analytical results.

4. Monitoring Locations: Surface Water Sites

Five sampling transects have been selected for surface water and sediment sampling in the lake. Each transect has two sites, yielding a total of 10 sites on the lake. When possible, each transect will have one littoral site and one limnetic site. Surface water samples will be collected at three depths at each site (surface, mid-depth, and bottom).

Three sites on Big Grizzly Creek have been selected for surface water sampling:
a) immediately upstream of the potassium permanganate detoxification station;
b) the 30-minute travel time mark downstream of the detoxification station; and
c) the 60-minute travel time mark downstream of the detoxification station.

5. Monitoring Locations: Other Sampling Sites

Five wells adjacent to or downstream of Lake Davis will be monitored. The campground well at the USFS Grasshopper Flat Campground will be sampled. Four wells on private properties will also be sampled; the locations of these four wells have been determined from a groundwater survey conducted by the Department of Water Resources.

The finished potable water at the plant outlet in the Plumas County Flood Control District water treatment plant will also be monitored.

6. Sampling Frequency

Sampling schedules have been developed for sampling rotenone in Lake Davis and Big Grizzly Creek (Table A); inerts in Lake Davis and Big Grizzly Creek (Table B); water quality in Lake Davis (Table C); rotenone and inerts in Lake Davis sediment (Table D); wells for rotenone, inerts, and bacteriological analysis (Table E); and water from the Plumas County Flood Control District Water Treatment Plant (Table F).C. Information Disclosure

A final report on the monitoring will be supplied to the Regional Water Quality Control Board-Central Valley Region, the California Department of Health Services, the Plumas County Health Department, and other interested parties within 60 days after the completion of monitoring.

D. Supervising Personnel

This study will be conducted by the CDFG Pesticide Investigations Unit under the supervision of Ms. Stella Siepmann. Mr. Brian Finlayson will be the primary contact person for other agencies and the public. All questions should be directed to him at (916) 358-2950.

E. References

American Public Health Association. 1995. Standard Methods for the Examination of Water and Wastewater. 19th Edition.

Dawson, V., P. Harmon, D. Schultz, and J. Allen. 1983. Rapid method for measuring rotenone in water at piscicidal concentration. Trans. Amer. Fish. Soc. 112:725-728.

Federal Register. 1987. Method 502.2; in: Environmental Protection Agency. Guideline establishing test procedures for the analysis of pollutants in drinking water.

Federal Register. 1992. Method 8010; in: Environmental Protection Agency. Guideline establishing test procedures for the analysis of pollutants in drinking water.

Federal Register. 1992. Method 8020; in: Environmental Protection Agency. Guideline establishing test procedures for the analysis of pollutants in drinking water.

Federal Register. 1992. Method 8310; in: Environmental Protection Agency. Guideline establishing test procedures for the analysis of pollutants in drinking water.

Harrington, J. and B. Finlayson. 1988. Rotenone residues in water following application to Kaweah River and Tulare Lake Basin, California. California Department of Fish and Game Environmental Services Division Administrative Report 88-1. Sacramento, California.

Sava, R. 1986. Guide to Sampling Air, Water, Soil and Vegetation for Chemical Analysis. California Department of Food and Agriculture. Sacramento, California.

II. Toxicological Monitoring

The purpose of toxicological monitoring is to (1) confirm the efficacy of the chemical treatment in the reservoir and tributaries, (2) to indicate the extent of downstream toxicity during the project, and (3) to indicate when the reservoir has detoxified to the extent that fish can be restocked. These objectives will be accomplished by the placement of live fish at various locations within the lake and downstream of the detox station during and after treatment.

A. Materials and Methods

1. Live cars will be placed at five of the chemical monitoring sites (one site on each transect) and five other locations on the lake, at the 15 minute and 30 minute sites on Big Grizzly Creek, and in all live tributaries upstream of the lake. Each live car set in the lake will consist of 3 cages attached to a weighted line suspended from a buoy. The highest cage will be suspended 1 meter from the surface, a cage will be at mid-depth, and the lowest cage will be approximately 1 meter from the bottom of the reservoir. The live cars in the streams will be single cages. Five rainbow trout will be placed in each cage.
2. In the lake, live cars will be set up before Day 0 and left in place until Day 21 (or later if conditions require). Fish will be checked daily. During the treatment, fish will not be replaced in lake live cars. Once fish at all lake locations die, the live cars will not be restocked with fish until Day 14. Fish will be replaced every 2 days if still alive.
3. Live cars in Big Grizzly Creek will be set up on Day 0 and checked three times daily. Dead fish will be replaced in the live cars used for detoxification. Fish will be replaced every 2 days if still alive.
4. Cages will be removed when fish survive and it is determined by chemical monitoring that the reservoir has detoxified.

III. Invertebrate Monitoring

This monitoring study is designed to determine the effects of rotenone on the recovery of aquatic invertebrate populations in Lake Davis.

The monitoring study will consist of two components: littoral macroinvertebrate sampling and plankton sampling. Sampling will take place in June, September, and October (posttreatment) of 1997 and in June and September of 1998.

A. Sampling Procedures

1. Littoral Macroinvertebrate Sampling

The littoral macroinvertebrate community will be sampled according to the California Lentic Bioassessment Procedures in productive shallow coves on the west side of the lake in July, September, and October (posttreatment) of 1997 and in June and September of 1998 and 1999. Ten wadable coves will be sampled by sweeping the substrate with a slack net. Five two to four meter transects will be sampled per cove. Contents of the net will be sieved and stored in 95% ethanol. The samples will be sent to the Water Pollution Control Laboratory for analysis.

2. Zooplankton Sampling

The zooplankton community will be sampled according to the California Lentic Bioassessment Procedures. Samples will be taken in productive shallow coves on the west side of the lake in July, September, and October (posttreatment) of 1997 and in

June and September of 1998 and 1999. Ten coves will be sampled from a boat. In each cove, a plankton net will be swept vertically at a set rate. The sample will be sieved and stored in 95% ethanol. The samples will be sent to the Water Pollution Control Laboratory for analysis.

B. Sample Analysis

Samples will be analyzed using California Stream Bioassessment Procedures—Macroinvertebrate Laboratory and Data Analyses. Upon arrival at the WPCL, sample jars will be opened and the contents evenly distributed in a gridded (5 cm^2) white enameled tray. Contents from randomly chosen grids will be used as subsamples. Macroinvertebrates from subsamples will be identified to the lowest possible taxonomic level. Taxonomic lists, diversity indices, and the following bioassessment metrics were generated for each sample: species richness (total number of genera and/or species present), Modified Hilsenhoff Biotic Index (family level tolerance values), percent contribution of dominant taxon (ratio of numerically dominant taxon to the total number of organisms), and EPT Index (total number of distinct taxa within the orders Ephemeroptera, Plecoptera, and Trichoptera).

C. Information Disclosure

A final report on the monitoring will be supplied to interested parties within 60 days completion of monitoring.

IV. TASTE AND ODOR MONITORING

Taste and odor of Lake Davis water will be tested before and after treatment to determine if treatment affects the potability of the water. Tragon Corporation has been contracted to perform this work.

A. Materials and Methods

1. In 1997, samples will be taken in June, August, October and twice in the month that the treatment plant is put back online. In 1998, samples will be taken in September.
2. Four water samples will be collected by CDFG from the water treatment plant and from three sites which receive water from Lake Davis during each sampling event. Two of these sites will be located in the City of Portola and one will be located in the Grizzly Lake Resort Improvement District. The samples will be collected in 1 gallon glass jugs.
3. Water samples will be transported to Redwood City and delivered to Tragon Corporation. Sample transport protocol (e.g., temperature) will be specified by Tragon. Within 72 hours of sample receipt, Tragon will use a panel of 25 screened consumers to evaluate the taste and odor of water from the treatment plant.
4. Tragon will provide CDFG with a report within five weeks of receiving the samples.

Sampling Schedule A Lake Davis and Big Grizzly Creek water for rotenone.

Site	Number of samples for rotenone, temperature, and dissolved oxygen by transect, site, and depth								
	Jun	Aug	Day -2	Day 2 (0–1)	Day 4 (3)	Day 7 (5–6)	Day 14 (8–13)	Day 21 (15–20)	Total
DL.A.1.1to3[1]	3	3	3	3	3	3	3	3	24
DL.A.2.1to3			3	3	3	3	3	3	18
DL.B.1.1to3			3	3	3	3	3	3	18
DL.B.2.1to3			3	3	3	3	3	3	18
DL.C.1.1to3	3	3	3	3	3	3	3	3	24
DL.C.2.1to3			3	3	3	3	3	3	18
DL.D.1.1to3			3	3	3	3	3	3	18
DL.D.2.1to3			3	3	3	3	3	3	18
DL.E.1.1to3	3	3	3	3	3	3	3	3	24
DL.E.2.1to3			3	3	3	3	3	3	18
BGC.1[2]			1	3(6)	3(3)	3(6)	3(18)	3(18)	67
BGC2			1	3(6)	3(3)	3(6)	3(18)	3(18)	67
BGC3			1	1(2)	1(1)	1(2)	1(6)	1(6)	22
Total	9	9	33	51	44	46	79	79	332

[1] DL=Davis Lake; A.=Transect A; A.1.=Site 1 of Transect A; A.1.1=Surface of Site 1 of Transect A; A.1.2=Mid-depth of Site 1 of Transect A; A.1.3=Bottom of Site 1 of Transect A.
[2] BGC=Big Grizzly Creek; 1=U/S detox station; 2=30 min D/S detox, 3=60 min D/S detox.

Sampling Schedule B Lake Davis and Big Grizzly Creek for inert compounds.

Site	Number of samples for EPA 502.2 and 8310 by transect, site, and depth								
	Jun	Aug	Day -2	Day 2	Day 4	Day 7	Day 14	Day 21	Total
DL.A.1.1to3[3]	3	3	3	3	3	3	3	3	24
DL.A.2.1to3			3	3	3	3	3	3	18
DL.C.1.1to3	3	3	3	3	3	3	3	3	24
DL.C.2.1to3			3	3	3	3	3	3	18
DL.E.1.1to3	3	3	3	3	3	3	3	3	24
DL.E.2.1to3			3	3	3	3	3	3	18
BGC2[4]			1	1	1	1	1	1	6
BGC3			1	1	1	1	1	1	6
Total	9	9	20	20	20	20	20	20	138

[3] DL=Davis Lake; A.=Transect A; A.1.=Site 1 of Transect A; A.1.1=Surface of Site 1 of Transect A; A.1.2=Mid-depth of Site 1 of Transect A; A.1.3=Bottom of Site 1 of Transect A.
[4] BGC=Big Grizzly Creek; 2= 30 min D/S detox, 3 = 60 min D/S detox.

Sampling Schedule C Lake Davis water for water quality (pH, BOD, alkalinity, hardness, total organic carbon, conductivity, and ammonia (or total nitrogen).

Site	Number of samples for water quality by transect and site				
	Aug	Day -2	Day 7	Day 21	Total
DL.A.1to3	3	3	3	3	12
DL.C.1to3	3	3	3	3	12
DL.E.1to3	3	3	3	3	12
Total	9	9	9	9	36

Sampling Schedule D Lake Davis sediment for rotenone and inerts.

Site	Number of sediment samples for rotenone and EPA 8310, 8010, and 8020 transect and site					
	Day -2	Day 2	Day 7	Day 14	Day 21	Total
DL.A.1	1	1	1	1	1	5
DL.B.1	1	1	1	1	1	5
DL.C.1	1	1	1	1	1	5
DL.D.1	1	1	1	1	1	5
DL.E.1	1	1	1	1	1	5
Total	5	5	5	5	5	25

Sampling Schedule E Wells surrounding Lake Davis for rotenone, inerts, and bacteriological analyses.

Site	Number of samples for rotenone , bacteriological, and EPA 502.2 and 8310 by Well									
	June	July	Aug	Day -2	Day 7	Day 21	Day 90	Day 180	Day 365	Total
Well 1	1	1	1	1	1	1	1	1	1	9
Well 2	1	1	1	1	1	1	1	1	1	9
Well 3	1	1	1	1	1	1	1	1	1	9
Well 4	1	1	1	1	1	1	1	1	1	9
Well 5	1	1	1	1	1	1	1	1	1	9
Total	5	5	5	5	5	5	5	5	5	45

Sampling Schedule F Collection of water from Plumas County Flood Control District Water Treatment Plant.

Analysis	Number of potable water samples for rotenone, bacteriological, BOD, EPA 502.2 and 8310 and taste/odor										
	Jun	Aug	Day -2	Day reuse	Day reuse +2	Day reuse +4	Day reuse +7	Day reuse +14	Day reuse 30	Day reuse +365	Total
Rotenone	1	1	1	1	1	1	1	1	1		9
502.2 & 8310	1	1	1	1	1	1	1	1	1		9
BOD	1	1	1	1	1	1	1	1	1		9
Bacteriological	1	1	1	1	1	1	1	1			9
Taste/Odor[5]	1	1	1		1			1		1	6
Total	5	5	5	4	5	4	4	5	4	1	43

[5] Samples will also be taken from three other locations in GLRID and City of Portola for taste and odor analysis.

APPENDIX F

SITE SAFETY PLAN

Lake Davis Northern Pike Eradication Project
Site Safety Plan
California Department of Fish and Game

This site safety plan is intended to identify and mitigate the potential safety hazards associated with the California Department of Fish and Game's (DFG) proposed plan to eradicate the exotic fish northern pike *Esox lucius* from Lake Davis, Plumas County, California. The eradication of northern pike will be accomplished by the application of two commercial formulations of the piscicide rotenone: Nusyn-Noxfish®, a liquid formulation and Pro-Noxfish®, a powder formulation. Nusyn-Noxfish® and Pro-Noxfish® are pesticide products that have been registered by the United States Environmental Protection Agency and the California Department of Pesticide Regulation. This plan is specifically designed for DFG employees and describes safety procedures and equipment that will be employed during each phase of the Lake Davis project including pretreatment preparations, the application of Nusyn-Noxfish® and Pro-Noxfish®, rotenone detoxification by potassium permanganate, monitoring, and posttreatment cleanup activities.

The DFG will use the incident command system (ICS) for coordinating project activities and for providing a central decision and communication base (Attachment 1).[1] The project safety officer (SO), as part of the ICS Command Team, has the primary authority to monitor and assess hazardous and/or unsafe situations and develop measures for assuring personnel safety. The SO will provide training sessions and conduct inspections of equipment and procedures used during the course of this project.

The potential hazards associated with the eradication of northern pike from Lake Davis can be categorized into the following five groups: (1) pesticide exposure, (2) non-pesticide chemical exposure, (3) spills, (4) general safety hazards, and (5) heat stress.

1. Pesticide Exposure

Title 3, Chapter 3, Sections 6700 through 6746 of the California Code of Regulations (CCR) describes the safety precautions required for employees that use pesticides. These regulations include provisions related to safety training, mixing and loading, pesticide application, and the handling and clean-up of pesticide containers and application equipment. The use of Nusyn-Noxfish® and Pro-Noxfish® by the California Department of Fish and Game will be accomplished in full compliance with these regulations, as summarized below. Compliance with state and federal pesticide safety regulations will be verified by inspections by the California Department of Pesticide Regulation and the Plumas County Agricultural Commissioner's Office.

A. Pesticide Application Supervision

The use of Nusyn-Noxfish® and Pro-Noxfish® will be supervised, on-site, by at least one DFG employee who has obtained his or her Qualified Applicator Certificate in the Aquatic Category from the California Department of Pesticide Regulation.

[1] Attachments 1–10 referenced in Appendix F are available from California Department of Fish and Game, 1701 Nimbus Road, Suite F, Rancho Cordova, California 95670, USA.

B. Pesticide Safety Training

All DFG employees involved with the Lake Davis Project will receive pesticide safety training specific to the use of Nusyn-Noxfish® and Pro-Noxfish®. This training will satisfy the requirements described in Title 3, CCR Section 6724 (Attachment 2), and the hazard communication (Attachment 3). A pesticide safety training record, DFG Form 1075, (Attachment 4) will be completed for each DFG employee who receives training.

C. Pesticide Safety Equipment and Procedures

All DFG employees that handle opened containers of Nusyn-Noxfish® and Pro-Noxfish® and/or participate in the application of these materials, are required to use the following safety equipment:

1) coveralls (disposable Tyvek®);
2) eye protection (splash goggles or full-face protection provided by the full-face respirators indicated below); and
3) nitrile gloves.

Additionally, employees must use respiratory protective equipment when handling Nusyn-Noxfish® and Pro-Noxfish®. Employees working with concentrated Nusyn-Noxfish® must use half-mask air-purifying respirators with organic-vapor removing cartridges with prefilters approved for pesticides (MSHA/NIOSH prefix TC-23C). Employees working with Pro-Noxfish® must use full-face respirators with dual cartridge fitted with dust filtering cartridges (MSHA/NIOSH prefix TC-21C) and organic-vapor removing cartridges (MSHA/NIOSH prefix TC-23C).

DFG employees that are required to wear respirators will be provided respiratory protection training that includes instruction on how to properly fit and test a respirator. Respirators will be assigned to each applicator by the SO. The use of respirators associated with rotenone use will be in full compliance with the regulations described in Title 3, Chapter 3, Section 6738(h) of the California Code of Regulations (Attachment 5).

All equipment must be clean and in good repair at the start of each work day. Each employee that handles Nusyn-Noxfish® and Pro-Noxfish® will be issued 2 sets of coveralls and gloves. Ripped or otherwise damaged equipment will be replaced as soon as possible. Extra sets of coveralls, gloves and eye protection (equal to approximately 50% of the total number of rotenone handlers) will be available in the event of equipment damage. Respirator cartridges must changed daily. Extra respirators and cartridges will also be available at the project site.

The exact specifications for required safety equipment can be found in Attachment 6.

D. Washing Facilities

Clean water, soap, and single-use towels for routine washing of the hands and face and emergency washing of the entire body will be available at the loading zone. An emergency eye wash station will be located at the loading zone.

E. Emergency Medical Care

An Emergency Medical Technician (EMT) will be on-site during the application portion of the project period. The EMT will be ready to respond to any medical emergencies, including those related to pesticide exposure. Additionally, the name, address and telephone number of a local physician, clinic, or hospital that can provide care in the event of a pesticide emergency will be posted in a prominent location at the Incident Command Post and at the loading zone.

F. Pesticide Emergencies or Illnesses

Employees that suspect that they have been made ill by the use of Nusyn-Noxfish® or Pro-Noxfish® must go immediately to the onsite EMT. The SO will be immediately notified of the situation. If necessary, the employee will be instructed to remove his or her work clothing, shower, and dress in clean clothing. In the event of Nusyn-Noxfish® or Pro-Noxfish® exposure to the eyes, the employee's eyes will be immediately flushed with large amounts of water for at least 15 minutes. The employee will be attended by the onsite EMT and, if necessary, will be taken to the appropriate medical care facility. Copies of the Nusyn-Noxfish® and Pro-Noxfish® product labels (Attachment 7) and material safety data sheets (Attachment 8) will be provided to both the onsite EMT and to any other medical professionals, as needed.

2. Non-pesticide Chemical Exposure

DFG will use the oxidant potassium permanganate to detoxify the rotenone-treated water in Big Grizzly Creek after its release from the outlet at Grizzly Valley Dam. Potassium permanganate is a strong oxidizer, non-volatile, non-flammable and stable under normal conditions. Hazardous exposures to potassium permanganate may occur via inhalation, ocular and dermal routes.

A. Personal Protective Equipment

All DFG employees involved in the use of potassium permanganate are required to use the following safety equipment:
1) coveralls (disposable Tyvek®);
2) eye protection (splash goggles); and
3) rubber or neoprene gloves.

Employees will be instructed to handle the compound in such a manner as to reduce the potential for potassium permanganate dust generation. If the possibility of overexposure to potassium permanganate exists, the use of a dust and mist respirator (NIOSH-MSHA TC-21C-287) will be required. Respirators will be available on site. DFG employees that are required to wear respirators will be provided respiratory protection training that includes instruction on how to properly fit and test a respirator.

All equipment must be clean and in good repair at the start of each work day. Clean safety equipment will be available on a daily basis for each employee that handles potassium permanganate. Ripped or otherwise damaged equipment will be replaced as soon as possible.

B. Washing Facilities

Clean water, soap, and towels for routine washing of the hands and face and emergency washing of the entire body will be available at the potassium permanganate detoxification station. An emergency eye wash station will also be located at the potassium permanganate detoxification station.

C. Emergency Medical Care

An Emergency Medical Technician (EMT) will be on-site during the application portion of the project period. The EMT will be ready to respond to any medical emergencies, including those related to chemical exposure. Additionally, the name, address, and telephone number of a local physician, clinic, or hospital that can provide care in the event of an emergency will be posted in a prominent location at the Incident Command Post and at the potassium permanganate detoxification station.

D. **Emergencies or Illnesses**

Employees that suspect that they have been injured or made ill by the use of potassium permanganate during the application portion of the project period must go immediately to the onsite EMT. The Project SO will be immediately notified of the situation. If necessary, the employee will be instructed to remove his or her work clothing, shower, and dress in clean clothing. In the event of exposure to the eyes, the employee's eyes will be immediately flushed with large amounts of water for at least 15 minutes. The employee will be attended by the on-site EMT and, if necessary, will be taken to the appropriate medical care facility. A copy of the potassium permanganate material safety data sheet (Attachment 9) will be provided to both the onsite EMT and to any other medical professionals, as needed.

3. Spills

The procedures related to Nusyn-Noxfish® and Pro-Noxfish® spill prevention and response are described in a separate document, the Northern Pike Eradication Spill Contingency Plan (Attachment 10).

Potassium permanganate spills should be cleaned up immediately by sweeping or shoveling up the material. Spilled materials should still be used in the rotenone detoxification process. Disposal of potassium permanganate wastes in landfills is prohibited.

4. General Safety Hazards

Among the general safety hazards that may be encountered during this project are slips, falls, improper lifting techniques, and heavy equipment or boating accidents. These potential safety hazards will be evaluated and mitigated via preventative actions.

The loading of Nusyn-Noxfish® and Pro-Noxfish® onto application boats will occur in the loading zone which will be posted as a "Hard Hat Use Area." No employees will be allowed within the loading zone without a hard hat and work coveralls. Forklift or crane operation zones will be posted and a general foot traffic restriction will be enforced in those zones.

Employees will be required to wear life vests at all times while on boats. Life vests will be worn in addition to the pesticide application safety gear described in Section 1(C).

Employees engaged in fish cleanup and disposal activities will be required to wear, at the minimum, cotton or Tyvek® disposable coveralls and disposable cotton or plastic gloves.

5. Heat Stress

Heat stress is a potential hazard to employees due to the requirements related to personal protective equipment. The wearing of moisture-impervious clothing is necessary to reduce employee exposure to Nusyn-Noxfish®, Pro-Noxfish®, and potassium permanganate. The use of moisture-impervious clothing, however, increases the employee's potential for heat stress by reducing the potential for body temperature cooling via sweat evaporation. All DFG personnel will be fully informed of the hazards related to heat stress. Further, DFG personnel will be required to drink 4-6 ounces of liquid during every 20 minutes of work, take appropriate work breaks and participate in heat stress monitoring when temperatures exceed 85°F.

Personnel that suspect that they are suffering from the effects of heat stress will be instructed to go immediately to the onsite EMT. The project safety officer will be immediately notified of cases of heat stress.

Approvals

____

Incident Commander Date

____

Safety Officer Date

____

Operations Section Chief Date

____

Staging Area Manager Date

____

Environmental Response Branch Supervisor Date

August 20, 1997

APPENDIX G

SPILL CONTINGENCY PLAN

Lake Davis Northern Pike Eradication Project Spill Contingency Plan California Department of Fish and Game

Description of Rotenone and Permanganate Products and Packaging

Two commercial formulations of rotenone will be used during the Lake Davis Project, Pro-Noxfish® (EPA Reg. No. 432-829) and Nusyn-Noxfish® (EPA Reg. No. 432-550). Approximately 130,000 pounds of Pro-Noxfish® and approximately 11,000 gallons of Nusyn-Noxfish® will be stored on-site at the Department of Water Resources (DWR) property adjacent to Big Grizzly Dam. The Department of Fish and Game (DFG) will take delivery of the rotenone at this site no earlier than October 1 and not later than October 6, 1997. The Pro-Noxfish® will be packaged in 200-pound fiber drums and the Nusyn-Noxfish® will be packaged in 55-gallon metal drums. Also present will be potassium permanganate used in the detoxification operation. Approximately 2,424 pounds (sufficient to treat 5 cfs at 3 mg/L for 30 days) will be present in 110- or 330-pound metal drums. Both rotenone formulations and the permanganate will remain in three or four tractor trailers used for shipment within the bermed storage area. Immediately prior to application the rotenone and the permanganate will be removed from the bermed area.

List of Materials:

1. (195) 55-gallon metal drums of Nusyn-Noxfish® (EPA Reg. No. 432-550) or 10,725 gallons;
2. (650) 200-pound fiber drums of Pro-Noxfish® (EPA Reg. No. 432-839) or 130,000 pounds; and
3. (24) 110-pound metal drums of potassium permanganate or 2,424 pounds.

Description of Storage Areas

The primary storage area which is adjacent to Big Grizzly Dam will be bermed to contain all of the liquid and powder rotenone and the potassium permanganate. The area will be graded so that drainage is to Lake Davis. The tractor trailers containing the rotenone will be locked and watched 24 hours a day. The bermed area will be large enough to contain all of the powder and the liquid inside the trailers. This will allow all of the rotenone to be recovered following a catastrophic spill. Total area minimum required is approximately 70 by 50 feet. The approximate 1,604 cubic feet of Nusyn-Noxfish®, 3,350 cubic feet of Pro-Noxfish®, and 50 cubic feet of permanganate will require approximately 5,000 cubic feet of containment. An area of 70 by 80 feet with a minimum of 12 inches of berm will provide 5,600 cubic feet of containment for the rotenone and permanganate. The primary storage area will be lined with heavy duty plastic; the perimeter will be enclosed with bales of hay (foundation for berm).

Portable bilge pumps, hoses, buckets, an empty tanker truck, potassium permanganate, shovels, extra bales of hay, clay, complete sets of safety gear, and absorbent pads will be maintained adjacent to the storage area from delivery date through the treatment date. At the time of treatment, chemicals will be transferred to the boat ramp, where they will be loaded onto boats for application.

A secondary storage area for the potassium permanganate will be constructed adjacent to the detoxification station just downstream of Big Grizzly Dam on Big Grizzly Creek. This storage area will be approximately 50 cubic feet and will be lined with heavy duty plastic; the perimeter will be enclosed with bales of hay (foundation for berm). This area will be utilized only during the detoxification operation.

Precautions

Only those pesticide containers which are in use will be opened at the application site. All personnel involved with the application will be knowledgeable of rotenone's toxicity, the Site Safety Plan, the rotenone and permanganate product labels, the material safety data sheets (MSDS), and the Spill Contingency Plan. Additionally, all personnel will wear proper clothing, eye protection, and respirators as specified in the Site Safety Plan.

In case of a spill, all personnel will have in possession the telephone numbers of the DFG Regional Manager, the DFG Pesticide Investigations Unit, the Plumas County Agricultural Commissioner, the Central Valley Regional Water Quality Control Board (Redding), California Department of Health Services (Redding), U.S. Forest Service Supervisor Office (Quincy) and Beckworth Ranger District, the State Office of Emergency Services, the Plumas County Health Department, the California Highway Patrol, the Plumas County Sheriff's Office, and the Chemtrec Hotline. All pesticides will be handled in accordance with the California Department of Pesticide Regulation (DPR) regulations and any other applicable regulations as specified in the Site Safety Plan.

Chain of Command

The flow of information and responsibility during application of rotenone to Lake Davis will follow the Incident Command System (ICS) structure (See Figure). The Environmental Response Branch, through the Staging Area Manager, through the Operations Section Chief will have the responsibility for immediate containment of a spill on-site (pending approval from the Safety Officer) during the application. For spills which occur prior to application, the Security Team will contact the Safety Officer, Liaison Officer, Information Officer, and the Incident Commander for appropriate cleanup instructions. All spills which present risk of upset to the environment are to be reported immediately to the Safety Officer, Liaison Officer, and the Incident Commander.

All mixing operations will be conducted on boats within the target area. Additionally, a DPR- Certified Applicator will be in charge of the Application Team and on the scene during mixing, loading, and application operations. A reconnaissance of the treatment area will be conducted by the person in charge of the application immediately prior to treatment.

A list of all known downstream water users on Big Grizzly Creek will be compiled and available at the storage site. For purposes of this contingency plan, the list will be limited to water users within ten miles downstream from Lake Davis.

Containment of Spills

In the event a spill occurs, it is of paramount importance that the spilled material be contained. Shovels and other hand tools will be used for immediate containment or channelization of the spilled material into a containment area. A spill of rotenone inside of the storage area will be recovered and used in the application. The following actions will be taken as necessary to contain a spill on ground:

1. Stopping the spillage at its source;
2. Diking in pools as appropriate;
3. Using materials such as clay, soil, sawdust, or straw to absorb standing material or collection of standing rotenone by pump or sponge and deposition into target area; and
4. Neutralizing the spill site with potassium permanganate as necessary.

Reporting Spills

Land spills of over 20 gallons of Nusyn-Noxfish® or 100 pounds of Pro-Noxfish® or potassium permanganate on the ground or any amount of rotenone below Grizzly Valley Dam into Big Grizzly Creek will be immediately reported to the following entities by the Incident Command (Safety Officer, Liaison Officer, and Information Officer):

1. DFG Regional Manager (Liaison Officer) Banky Curtis (916) 358-2899

2. DFG Pesticide Investigations Unit Supervisor (Incident Commander) Brian Finlayson (916) 358-2950

3. Plumas County Agricultural Commissioner Carl Bishop (916) 283-6365

4. Plumas National Forest Supervisor Mark Madrid (916) 283-2050

5. Plumas National Forest Mohawk District Ranger (916) 836-2575

6. Plumas County Environmental Health Director William Crigler (916) 283-6355.

7. California Regional Water Quality Control Board Engineer George Day (916) 224-4854

*8. California Department of Health Services District Engineer Gunther Sturm (916) 224-4866

*9. California Office of Emergency Services Warning Center 1(800) 852-7550

10. Chemtrec Hotline (800) 424-9300

*12. California Highway Patrol (916) 445- 2211

*13. Plumas County Sheriff's Office (916) 283-6300

*14. Plumas County Office of Emergency Services Director Andy Anderson (916) 283-6268

For small land spills (under 20 gallons or 100 pounds) which do not threaten the environment, reporting can be made at the earliest convenience of the applicator and can be restricted to the first six persons on the above list. In the event of a major spill (defined by the amount of rotenone in receiving water that could exceed the DHS Action Level of 4 g/L rotenone in a potable water supply) into water outside the project area or onto land which can pass into water outside the project area, all downstream potable water supplies within ten miles will be immediately notified by DFG personnel with assistance from government agencies (marked with * above), as necessary.

Spill Treatment

The rotenone which has been spilled and absorbed into the dirt should be removed and applied to the application area. The contaminated dirt will be treated as if it where the rotenone pesticide, and all required pesticide application safety gear will be worn. If there is a spill outside of the project area and there is a chance that the contaminated soil could wash into surface water out of the project area, then the disposal of the contaminated soil to an approved landfill may be required. The permanganate which has been spilled will be recovered and used, where possible. Disposal of spilled material and contaminated soil will be made in accordance with requirements of the California Regional Water Quality Control Board (see number 7 on Report Spills). The spilled material should be put into the application site (Lake Davis), where possible. All materials in contact with the rotenone and potassium permanganate will be washed in an area adjacent to the reservoir so that all rinsate will flow into project waters.

Approvals

____
Incident Commander Date

____
Safety Officer Date

____
Operations Section Chief Date

____
Staging Area Manager Date

____
Environmental Response Branch Supervisor Date

TECHNICAL PROCEDURES 3

Rotenone is available in two liquid formulations (5% active ingredient as Noxfish® or 2.5% active ingredient with 2.5% synergist as Nusyn-Noxfish®) and one powder formulation (generally 5% or greater active ingredient as Pro-Noxfish®). These formulations are available from AgrEvo and are used only for discussion purposes in this manual; other similar products (based on type and percent active ingredient) are available from other manufacturers, such as Prentiss, and can be used interchangeably with the referenced products (see Appendix H). Treatment concentrations of 0.005–0.250 ppm rotenone (0.1–5 ppm formulation) are recommended on AgrEvo labels for Noxfish® and Pro-Noxfish®. The Nusyn-Noxfish® (USEPA approved) label recommends the same treatment concentrations of rotenone, but twice (0.2–10 ppm) the concentration of the formulation due to the percentage of active ingredient in the product (Table 3.1a). These USEPA instructions do not acknowledge the effect of the synergist that doubles the toxicity of rotenone to fish; however, this is acknowledged on the Canadian (PMRA approved) label for Nusyn-Noxfish® (Table 3.1b). Regardless, most biologists apply Nusyn-Noxfish® at the same rate as Noxfish® or one-half the rotenone concentration (i.e., 1 ppm Nusyn-Noxfish® and 1 ppm Noxfish® have similar effects). This is not a violation of the label because concentrations of rotenone are within label allowance. Bioassays (see Appendix I) using water and fish from the treatment project water will indicate if the intended concentration is lethal.

3.1 TREATMENT OF PONDS, LAKES, AND RESERVOIRS

Ponds, lakes, and reservoirs are characterized by size, depth, and water quality. Size is an important determinant of effort and planning required for a rotenone treatment even though similar procedures are followed on all standing waters. Water quality parameters that include temperature, pH, alkalinity, algae, organic content, and sunlight penetration influence the toxicity and rate of natural degradation of rotenone.

Rotenone suppliers (see Section 1.5) generally describe applications based on the following uses: (1) selective treatment; (2) normal pond use; (3) removal of bullheads or carp; (4) removal of bullheads or carp in rich organic ponds; and (5) preimpoundment treatment above dams.

Table 3.1a. Application rates and concentrations of rotenone needed to control fish (from USEPA approved labels) in lakes, ponds, and reservoirs. This table indicates that twice the amount of Nusyn-Noxfish® compared to Noxfish® is required. However, most use Nusyn-Noxfish® at the same rate as Noxfish®. This practice does not violate the label requirements because Nusyn-Noxfish® is used at a lower concentration than the label allows. (Source: AgrEvo labels which are in Appendix H, AF = acre-feet).

Type of use	Active rotenone (ppm)	Noxfish®		Nusyn-Noxfish®		Pro-Noxfish®	
		ppm	AF/gal	ppm	AF/gal	ppm	AF/lb
Selective treatment	0.005–0.007	0.10–0.13	34–24	0.20–0.25	15–12	0.10–0.13	3.7–2.8
Normal pond use	0.025–0.050	0.5–1.0	6.0–3.0	1.0–2.0	3.0–1.5	0.5–1.0	0.74–0.37
Remove bullheads and carp	0.050–0.100	1.0–2.0	3.0–1.5	2.0–4.0	1.5–0.75	1.0–2.0	0.37–0.185
Remove bullheads and carp in organic ponds	0.100–0.200	2.0–4.0	1.5–0.75	4.0–8.0	0.75–0.38	2.0–4.0	0.185–0.093
Preimpoundment treatment above dam	0.150–0.250	3.0–5.0	1.0–0.60	6.0–10.0	0.50–0.30	3.0–5.0	0.123–0.074

Table 3.1b. Metric equivalents of application rates and concentrations of rotenone needed to control fish (from Canadian PMRA approved labels) in lakes, ponds, and reservoirs. Note that m^3/L units are used for the applications instead of ha-m/L. (Source: AgrEvo labels, which are in Appendix H and W. Stetter, Foreign Domestic Chemicals Corporation, personal communication, 1999).

Type of use	Active rotenone (ppm)*	Noxfish®		Nusyn-Noxfish®		Cube powder	
		ppm	m^3/L	ppm	m^3/L	ppm	m^3/kg
Selective treatment	0.005–0.007	0.10–0.13	9,777–7,821	0.10–0.13	9,777–7,821	0.10–0.13	10,060–7,610
Normal pond use	0.025–0.050	0.5–1.0	1,955–978	0.5–1.0	1,955–978	0.5–1.0	2,020–1,010
Remove bullheads and carp	0.050–0.100	1.0–2.0	978–489	1.0–2.0	978–489	1.0–2.0	1,010–505
Remove bullheads and carp in organic ponds	0.100–0.200	2.0–4.0	489–244	2.0–4.0	489–244	2.0–4.0	505–250
Preimpoundment treatment above dam	0.150–0.250	3.0–5.0	325–196	3.0–5.0	325–196	3.0–5.0	330–200

* Multiply by 0.5 for the active rotenone (ppm) from Nusyn-Noxfish®.

Treatments of standing waters are further classified as selective, partial or total removal, or sampling. The objective of a selective treatment is to remove or reduce the population of a selected species by treating the entire water body with a concentration of rotenone that targets a particular life stage or species (e.g., remove or reduce the size of a bluegill population). A partial treatment has similar objectives, but targets only specific areas of a body of water for removal of all fish. The objective of a total removal is to eliminate all fish from the entire water body. Rotenone is used in many states to sample fish populations by treating shorelines, coves, or stream reaches. These areas are blocked with nets to prevent escapement and to facilitate collection of fish. The most common use of rotenone in North America is for maintenance of sport fisheries; the second most common use is to quantify fish populations (see Appendix A).

Regardless of the project's objective, the formulation of rotenone that will be applied must be determined. Liquid formulations are easier to apply and disperse effectively in heavily vegetated areas of standing water. Liquid rotenone is the only effective formulation for treating flowing water. Emulsifiers, solvents, and synergists found in liquid formulations cause water quality and public health concerns not associated with powder (see Section 5). The development of a rotenone aspirator by the Utah Division of Wildlife Resources (UDWR) in 1990 provided a more efficient method of applying powdered rotenone to standing water. High pressure pumps used to operate the powder aspirators are more expensive than the gravity feed or trash pumps used for application of liquid formulations. However, the cost of powder formulations is about one-third the cost of liquid formulations.

3.1.1 Calculation of water volume and rotenone requirement

The volume of water can be determined with bathymetric maps or by mapping a pond, lake, or reservoir (standing water) and determining volume. There are a variety of methods available to calculate water volume (Gallagher 1999). Agencies that control standing waters generally have volume information. A global positioning system (GPS) instrument or a plane table and planimeter can be used to map shorelines. Depths can be measured along transects on the water using a measuring rod or sounding line, and marking positions with the GPS or planimeter. The surface area and volume can then be determined for the contours from the map that is created. The volume of water is calculated in acre-feet (or m^3). For large bodies of water, bathymetric maps can be digitized to determine volume at each contour. It is recommended that large lakes or reservoirs be segmented into smaller sections and each section treated separately.

All tributaries to standing waters should be identified and mapped. In some projects, it may be desirable to treat only the segment of the tributary immediately upstream from the reservoir to prevent escapement of fish during the treatment. In other cases, treatment of tributaries will be part of the total treatment project to ensure that undesirable fish will not be left in the system. Application of rotenone to streams should begin immediately before the treatment of the standing water and continue throughout the treatment. Refer to Section 3.2 for procedures used in flowing waters.

The degradation rate of rotenone, which influences its effectiveness, is affected primarily by temperature and sunlight. Gilderhus et al. (1986) calculated the half-life of rotenone at 13.9 and 83.9 h in ponds at temperatures of 24°C and 0°C, respectively. Finlayson and J. Harrington (unpublished data, presented at Chemical Rehabilitation Projects Symposium, Bozeman, Montana, 1991) reported the half-life of rotenone at 41.8 and 84 h in reservoirs at temperatures of 20–22°C and 10–20°C, respectively. Alkalinity and pH also influence rotenone degradation. Finlayson and Harrington (unpublished data, 1991) reported that waters with high alkalinity (>170 ppm $CaCO_3$) and pH (>9.0) degraded rotenone faster than waters with lower alkalinity and pH. Degradation in summer is accelerated by both higher water temperature and greater sunlight exposure. Additional rotenone is required in waters with high temperature, pH, alkalinity, and sunlight penetration. Organically rich waters with high volumes of suspended solids and algae will also require higher concentrations of rotenone; if this is a concern, determine the effectiveness of the rotenone formulation in the water to be treated by bioassay (see Section 3.1.1).

To determine the desired concentration of chemical to use for the project, the physical attributes of the water body and species of fish should be compared to the recommendations on the product label. Marking and Bills (1976) present lethal concentration and 50% mortality values (LC50) for 24 and 96 h for a great variety of fish species (see Appendix I). Goldfish *Carassius auratus* and black bullhead *Ameiurus melas* had the greatest tolerance to rotenone. Both species were 10 times more tolerant than most other species. The label indicates that bullhead and carp may be treated at concentrations of 0.050–0.200 ppm rotenone. Marking and Bills (1976) reported 96 h LC50 values for goldfish and black bullhead of 0.025 ppm and 0.019 ppm rotenone, respectively, under laboratory conditions.

The product labels for rotenone formulations in Table 3.1 specify the number of acre-feet of water that will be treated with the particular formulation. The amount of powder formulation needed is based on 5% active ingredient; however the concentration of rotenone may vary by lot. The amount of powder formulation needed must be adjusted to reflect the actual percentage of active ingredient. To determine this, multiply the number of pounds recommended for the desired concentration listed in Table 3.1 by the fraction (5%/lot%). For example to treat at 0.050 ppm with Pro-Noxfish® that is 7% rotenone, multiply the lb of powder formulation needed by 0.714 (5%/7% = 0.714).

To determine the lb of powder or gal of liquid rotenone from Table 3.1a, use the formulas below:

1. Pounds of Pro-Noxfish® needed = Volume of the impoundment in AF divided by the AF per lb of Pro-Noxfish® needed for the desired concentration, and;
2. Gallons of Noxfish® needed = Volume of the impoundment in AF divided by the AF per gal of Noxfish® for the desired concentration.

Although these formulas are taken from AgrEvo labels they will generally be applicable to other supplier's rotenone products.

General formulas when the table is not used are:

1. Pounds of 5% powder needed = Volume of impoundment in AF x 2.72 lb x ppm desired concentration.
2. Gallons of 5% liquid needed = Volume of impoundment in AF x 0.33 x ppm of desired concentration.

An AF of water contains 325,872 gal, which weighs 2,720,000 lb. To provide one ppm of active ingredient in 1 AF of water, 2.72 lb of 5% powder or 0.33 gal of 5% liquid rotenone is required.

Bioassays before treatment will indicate if the recommended concentration is effective. Careful consideration should be given to the use of bioassays to determine required concentrations. When total removal of fish is the objective, it may be advisable to treat at concentrations higher than the minimum effective dose demonstrated by the bioassay.

Bioassay tests should be conducted using target fish species, rotenone from the stock to be used, and water from the body of water that will be treated to determine if the actual concentration needed for a particular application will work satisfactorily (for bioassay techniques see Appendix J). The concentration applied must be no higher than the recommended concentrations on the label to be in compliance with U.S. Environmental Protection Agency (USEPA) regulations. Some labels may be more restrictive than others. The California Department of Fish and Game (CDFG 1994) generally treats streams with Nuysn-Noxfish® at a concentration of 1.0 ppm and lakes at 2.0 ppm. Expected active rotenone concentrations are approximately 0.025–0.050 ppm following complete mixing. However, many states treat at the highest rate allowed on the product label to ensure that a complete a kill is achieved. In highly alkaline waters in midwestern states it is not uncommon to use 4–5 ppm of 5% formulations (G. Tichacek, retired, Illinois Department of Natural Resources, personal communication, 1999). Other states and provinces may have more restrictive requirements than the label instructions. For instance, New York Pesticide Regulations restrict treatments to a maximum of 1.0 ppm rotenone formulation (T. Nashett, New York State Department of Environmental Conservation [NYDEC], personal communication, 1999).

Bioassays can result in chemical and manpower savings if lower concentrations are demonstrated to be effective. For example, UDWR (C. Thompson, UDWR, personal communication, 1999) found that carp could be effectively killed in organically rich water at a concentration of 1.5 ppm Pro-Noxfish®, which is less than the label recommendation of 2.0–4.0 ppm. This resulted in a substantial savings in chemical costs and application time.

3.1.2 Treatment techniques and equipment

Maps of large standing waters can be digitized and divided into sections of known volume that can be treated in one to two days. It is advisable to complete the project within 48 h so that the entire water body is at the desired concentration before the rotenone degrades significantly. This timing may not be possible on very large projects. However, if the treatment is done by systematically moving applications in one direction, fish should remain in toxic water.

Sections should be further divided into subsections and assigned to individual boat operators with the appropriate amount of rotenone to apply. Marker flags or buoys are used to delineate the sections. The boat operator and project manager record the amount of rotenone applied. Ponds and small reservoirs are usually treated from shore or from one boat.

When sampling of fish populations is the objective, the amount of rotenone needed and the application techniques are the same as those used for total or partial treatments. Davies and Shelton (1983) describe techniques and biases associated with this method of population and stock assessment. Biases not usually associated with the application of rotenone include site selection, sample design, incomplete recovery of fish, and distribution patterns.

3.1.2.1 Liquid applications

The equipment needed to treat standing water depends on area, depth, and location. Bodies of water that cannot be sprayed from shore with conventional commercial pesticide sprayers require motorized boats to distribute rotenone. If more than 200 gal of rotenone is applied by a boat in a 2-day period, the use of large, gas-powered pumps may be desirable (Figure 3.1a and 3.1b). See Appendix K for additional information about equipment and techniques. A liquid formulation can be applied using a venturi boat bailer system (Figure 3.1c). Rotenone is siphoned from the container through a hose to a venturi apparatus attached to the motor housing or hull of the boat. The Illinois Department of Natural Resources has treated lakes up to 600 surface acres using venturi boat bailers (Tichacek, personal communication, 1999).

Commercial pesticide sprayers are used for smaller bodies of water. Sprayers are hand pumped, electric, or gasoline powered. Typical power sprayers are equipped with 10–100-gal pesticide tanks and are mounted on pickups, all-terrain vehicles, or in small boats.

An even distribution of rotenone in each section of the body of water is critical for an effective removal of target fish. The liquid formulation must be diluted 1: 10 with water if it is removed from the original container. Equipment now available allows for the automatic dilution of the product with water prior to application, thus allowing for a closed-system applica-

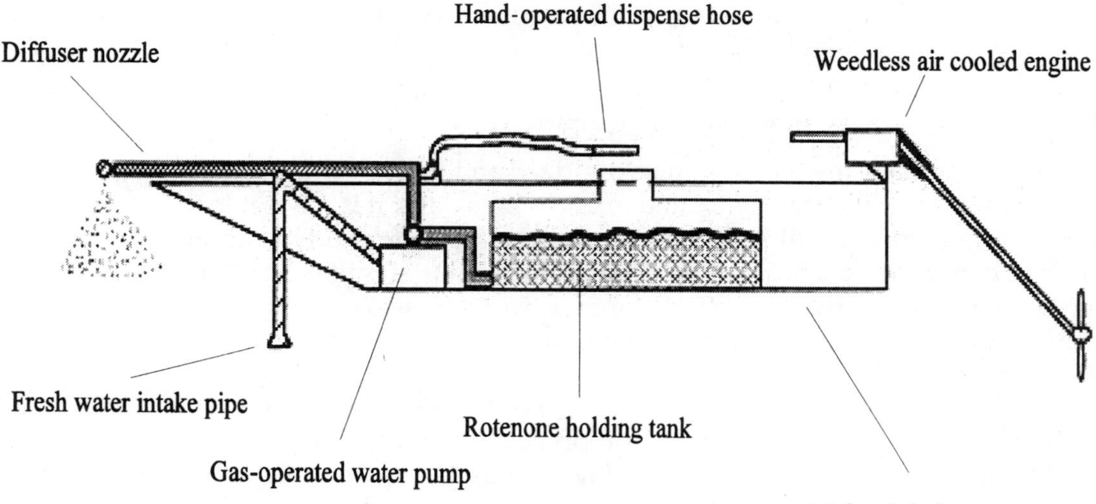

Figure 3.1a Johnboat adapted for distribution of liquid rotenone by Utah Division of Wildlife Resources. Operated by an air-cooled, shallow draft engine, with 50-gal rotenone tank and water pump to draw water from the lake and spray onto the water surface. Rotenone is fed into the system by venturi or gravity through a valve in the tank.

Figure 3.1b Liquid rotenone dispensing boats with rotenone container, pump, suction hose, hand sprayer, and diffuser.

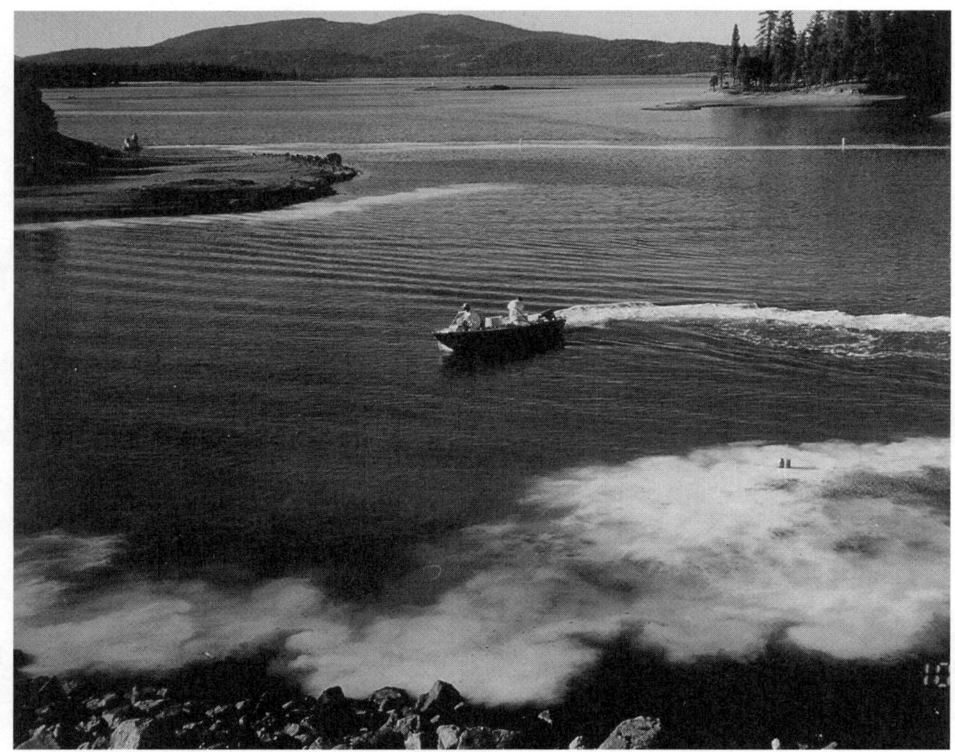

Figure 3.1c Liquid rotenone dispersion using a boat with a holding tank and boat bailer system. Lower figure shows rotenone being dispensed with this equipment.

tion where applicator contract with the concentrated product is avoided (see Appendix K). This dilution is desirable for a more uniform distribution when rotenone is drawn directly from the container, mixed, and applied to the water through a pump siphon or venturi system. Liquid formulations disperse rapidly horizontally and vertically in shallow impoundments. Both liquid and powdered formulations are heavier than water and will sink. Simple repetitive dispersal patterns throughout each segment will evenly distribute the rotenone. Surface application rates may need to be increased to account for suspended sediment, organic matter, or extensive vegetative growth, each of which tends to reduce the effectiveness of rotenone.

It may be necessary to pump rotenone into deep water when a strong thermocline exists. Many jurisdictions have dealt with this situation by employing extended discharge hoses weighted with heavy trolling weights to prevent the hose from surfacing. Vertical mixing may be further facilitated by extending the water pump suction line near the bottom to draw cold, dense water to the surface where the rotenone is mixed. The NYDEC has found it effective to treat stratified lakes at 10-ft depth intervals (L. DeMong, NYDEC, personal communication, 1999). The rotenone is pumped into each 10-ft interval of depth to apply the treatment concentration based on the calculated volume of water in the interval.

The injection of rotenone into deep water is only necessary when rotenone concentrations in deep areas are insufficient for a complete kill two days after application. Problems have occurred where rotenone was injected at depth adjacent to a discharge structure on a dam. Rotenone discharged through the dam killed fish in downstream areas for several miles (CDFG 1994). Surface application at the maximum label rates gives the formulation time to defuse throughout the water column and may eliminate the need to treat deeper layers. It may also be desirable to complete treatments while the lake is stratified if oxygen concentrations are low below the thermocline. In such cases, target fish may not survive in deep water, and the treatment of a smaller volume of water above the thermocline may be successful. The extent of oxygen depletion must be carefully determined. Some states and provinces have experienced failed treatments by assuming dissolved oxygen concentrations were too low to support fish below the thermocline.

Deep lakes have been effectively treated in Michigan and Minnesota just before ice cover, using lower concentrations of rotenone. Rotenone remains toxic longer in cold water, which provides longer exposure time. This technique has been effective for removing resistant species (e.g., bullheads) from lakes. The concentrations of rotenone have remained toxic in some lakes for two or more months.

For aerial treatments, large droplets or streams of dilute rotenone are preferred over mist or small droplet applications. Mist or small droplet applications may result in drift that reduces treatment efficacy and increases the risk of detrimental effects on nontarget organisms.

3.1.2.2 Powder application

An aspirator system for powder applications was developed by UDWR to create a slurry of powder and water for the renovation of Strawberry Reservoir in 1990. The rotenone aspirator was made with galvanized pipe and fittings available from most plumbing supply stores (Figure 3.2; also see Appendix L). Approximately 880,000 lb of powdered rotenone was applied to the reservoir in five days with four 30-ft barges and two National Guard ribbon bridges moved about the lake by tender boats (Figure 3.3a and 3.3c). This system was modified by the North Dakota Department of Game and Fish, the Jicarilla Apache Tribe, CDFG, and UDWR through the use of 16- and 18-ft johnboats and platform boats (Figure 3.3b).

The rotenone aspirator nearly eliminates dust associated with powder applications. An efficient crew can vacuum nearly 100% of the powder from the container at a rate of 20–40 lb per min. Empty rotenone containers

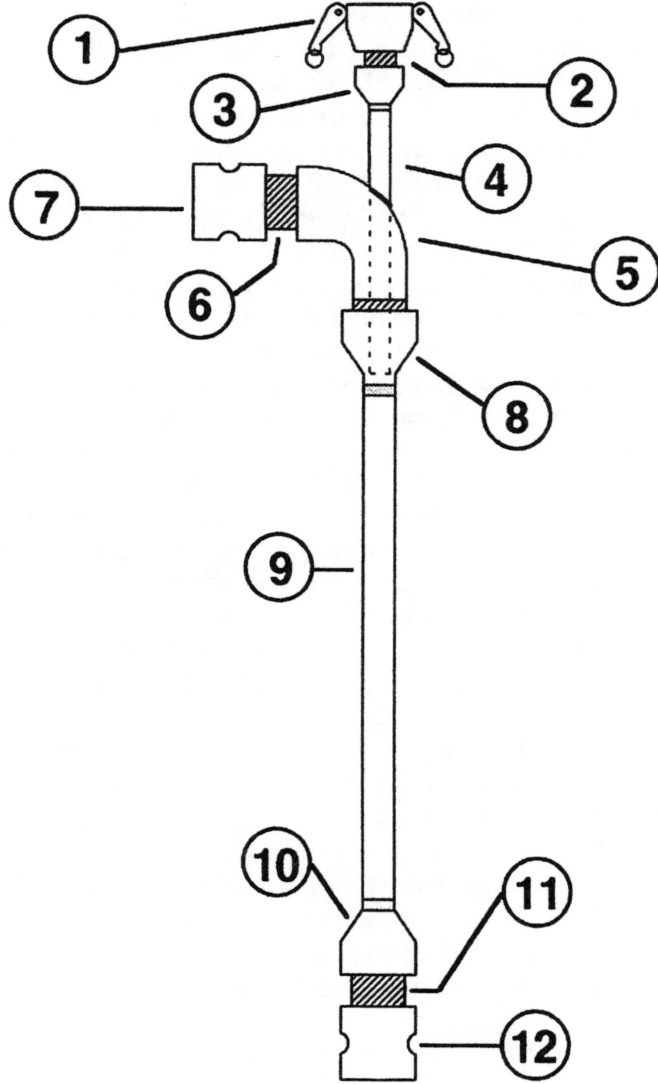

Figure 3.2 Rotenone aspirator. 1) 2-in quick release fitting; 2) nipple; 3) reducer; 4) 1.25-in suction pipe; 5) 3-in street elbow; 6) nipple; 7) 3-in quick release fitting; 8) bell housing; 9) pipe; 10) reducer; 11) nipple; 12) 3-in quick release fitting.

and liners are disposed of according to local ordinances. The pesticide label (see Appendix H) requires disposal of the container in a sanitary landfill or by incineration.

3.2 TREATMENT OF STREAMS AND RIVERS

Flowing water is characterized by depth, width, and velocity. All treated streams must be carefully surveyed and mapped. Water flow in cubic feet per second (ft^3/s) and velocity in feet per second (ft/s) must be measured throughout the treatment zone to assess fluctuations in discharge. A variety

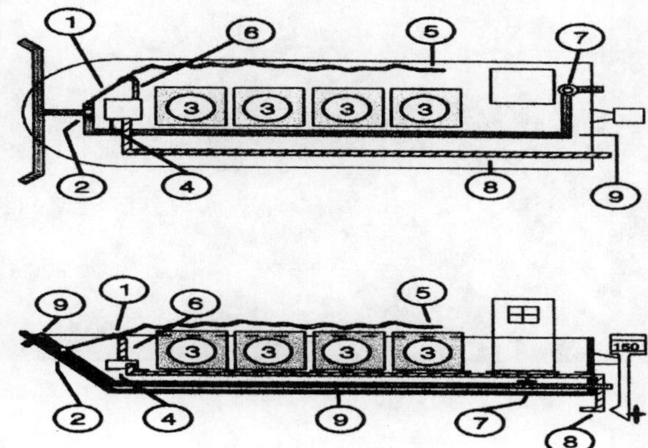

Figure 3.3a Powder slurry mixing barges developed by Utah Division of Wildlife Resources. Barge is 30 ft long powered by a 150 hp outboard motor. 1) rotenone aspirator; 2) gate valve; 3) 1,000-lb bulk bags; 4) high pressure pump; 5) vacuum hose and PVC pipe; 6) high pressure hose; 7) gate valve; 8) water delivery hose; 9) slurry discharge hose.

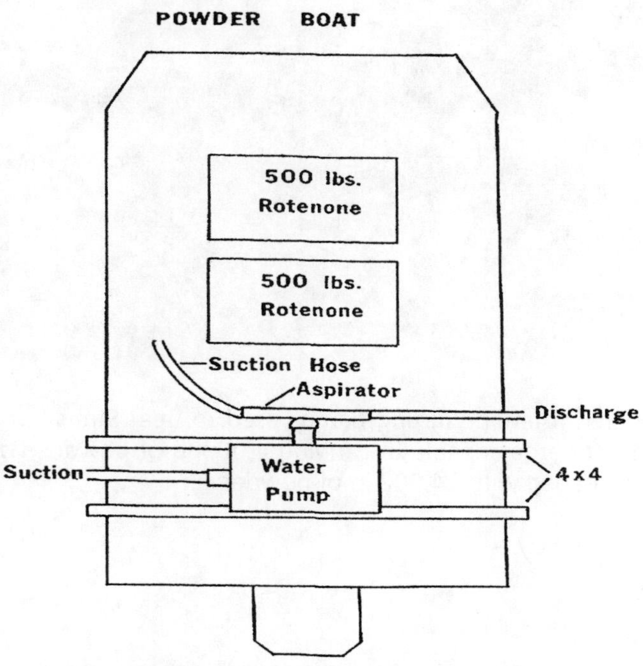

Figure 3.3b Johnboat containing 500-lb bags of powdered rotenone and aspirator mixer (developed by Tom Watts, Game and Fish Department, Jicarilla Apache Tribe).

Figure 3.3c Powdered rotenone mixing barges used to treat Strawberry Reservoir, Utah, 1990. Upper figure is a 30-ft barge capable of carrying 4,000 lb of powder. Lower figure is a National Guard ribbon bridge loaded with 24,000 lb of powder.

of methods are available to measure steamflow (Gallagher and Stevenson 1999). Water volume and velocity in streams change constantly depending on precipitation in the watershed and groundwater recharge. Water volume and velocity should be measured in advance of the project to estimate the amount of rotenone necessary and then remeasured immediately before the project. Stream treatments may be associated with treatment of standing water. It is necessary to treat tributaries to impoundments to ensure that the target species do not avoid treatment by migrating to untreated waters. Treatment of the tributaries begins before the treatment of an impoundment.

Barriers, beaver dams, seeps, springs, and tributaries are identified and mapped in advance of the treatment. Barriers such as diversion structures and beaver ponds should be removed, if possible, before the treatment. If they cannot be removed, their rotenone demand must be carefully measured. If barriers cannot be removed and water is stored for more than one to two hours, it may be advisable to set drip stations below such barriers to boost rotenone concentrations.

Similar procedures are followed on any size stream. Depending on the length and volume of the stream, one or more drip stations will be required to apply and maintain the target concentration (see Section 3.2.2). If the goal of the treatment is total removal of all fish, multiple treatments in the same or successive years may be required.

3.2.1 Calculation of amount of product and active chemical ingredient

The chemical requirement for treatment of streams depends on discharge, target species, and the duration of the treatment (refer to Appendix M, Noxfish® and Nusyn-Noxfish® Stream and River Use Monographs, AgrEvo, undated). AgrEvo, one rotenone supplier currently with a pesticide label for solely treating flowing waters (e.g., flowing water treatments not associated with treatments of ponds, lakes, and reservoirs) recommends that slow-moving rivers with little or no water exchange be treated as impoundments. AgrEvo recommends rotenone application for 4–8 h to achieve a complete kill in streams. The Michigan Department of Natural Resources (W. McClay, Michigan Department of Natural Resources, personal communication, 1999) recommends a 30-min contact time when using rotenone as a sampling tool. Multiple application sites may be needed because rotenone is diluted with nontreated water and detoxified by sunlight over distance. AgrEvo recommends that sites be separated about ½ to 2 miles apart, or not more than 2 h or less than 1 h travel time (see Appendix N).

If it is not possible to place drip stations between 1 to 2 h travel time apart, we recommend that the treatment duration be long enough to provide for four complete volume turnovers of water between the drip stations. For example, a stream stretch (between two drip stations) of 5,280 ft by 20 ft wide by 1 ft deep has a volume of 105,600 ft^3. A water flow of 10 ft^3/s through the stream stretch yields a volume of 36,000 ft^3/h. One complete volume change (turnover) in this stream stretch would occur in (105,600 ft^3/36,000 ft^3/h) = 2.9 h. Thus, the recommended treatment time for this stream stretch (2.9 h/turnover x 4 turnovers) is 11.6 h.

The time for water movement between two sites is best determined using fluorescent dye. Dye such as uranine or rhodamine WT is detected visually or with a fluorometer at nonvisual levels. Automatic water samplers may be used to periodically collect samples at sites along the length of the stream.

For flows less than 25 ft^3/s the Stream and River Use Monographs (see Appendix M) state that the liquid formulation should be injected directly into the center of the stream at a rate of 0.85–3.4 mL/min for each ft^3/s of stream discharge. This is equivalent to 0.5–2.0 ppm Noxfish® (0.025 to 0.100 ppm rotenone) and from 0.5 to 2.0 ppm Nusyn-Noxfish® (0.012 to 0.050 ppm rotenone). In large streams, it is advisable to apply the rotenone across the stream to assure good distribution. This is especially true when fish population sampling in short-stream reaches is the objective of the project.

Bioassays should be completed on target species in the streams before the treatment to determine the concentration of rotenone needed to kill the target species. In small headwater streams, 0.5 ppm liquid formulation or less will effectively remove trout. However, if these streams are very cold, higher concentrations will be necessary. In large slow moving rivers, 2.0 ppm liquid formulation may be required.

The equations for determining application rates for Noxfish® and Nuysn-Noxfish® are found in the Stream and River Use Monographs (see Appendix M).

The CDFG has developed a 5-gal drip head system that can be adjusted to dispense rotenone for 4–8 h (Figure 3.4a and 3.4b). Table 3.2 is an application chart for using an 79, 53, and 39 mL/min drip head for 4, 6, and 8 h, respectively.

The UDWR uses a drip head system in which a 1/8- to 1/16-in hole is drilled in the bottom of a "T" fitted with a metal plug and attached to a barrel (Figure 3.5). In this system the continuous drip through the drip head does not dispense at a rate of 80 mL/min, but must be calibrated for each drip head. The volume of water and rotenone that must be applied to the stream for the duration of the treatment can be calculated with the following equation:

Table 3.2. Chart for the application of liquid rotenone at 1.0 ppm to flowing water using a 5-gal reservoir for the drip head system. To increase the treatment rate to 2.0 ppm, increase the amount of liquid rotenone by 2-fold (decrease water by an appropriate amount); to decrease the treatment rate to 0.5 ppm, decrease the amount of liquid rotenone by one-half (increase water by an appropriate amount). Values in gallons.

	4-h treatment (79 mL/min)		6-h treatment (53 mL/min)		8-h treatment (39 mL/min)	
ft^3/s	Rotenone	Water	Rotenone	Water	Rotenone	Water
1	0.108	4.892	0.162	4.838	0.216	4.784
2	0.216	4.784	0.323	4.677	0.431	4.569
3	0.323	4.677	0.485	4.515	0.647	4.353
4	0.431	4.569	0.647	4.353	0.862	4.138
5	0.539	4.461	0.808	4.192	1.078	3.922
6	0.647	4.353	0.970	4.030	1.294	3.706
7	0.755	4.245	1.132	3.868	1.509	3.491
8	0.862	4.138	1.294	3.706	1.725	3.275
9	0.970	4.030	1.455	3.545	1.940	3.060
10	1.078	3.922	1.617	3.383	2.156	2.844

Figure 3.4a Biologist in safety gear calibrating the delivery of rotenone from a continuous drip can into a stream.

Figure 3.4b Close up showing plumbing of continuous drip can. (Note: This can is empty. Appropriate safety gear must be worn when applying rotenone.)

$$\text{Total gal of chemical} = \text{ppm desired concentration} \times 450 \text{ gal/min} \times \text{waterflow as ft}^3\text{/s} \times \text{min applied}/1{,}000{,}000$$

(1 ft^3/s = 450 gal/min).

Determine the number of hours of treatment (4, 6, or 8) and the intended treatment concentration (1 or 2 ppm). Determine the rate of flow through the drip head (mL/min), the amount of liquid formulation needed for the desired concentration for the time of treatment (formula above), and the amount of water and rotenone (diluted rotenone) that will drip during the time period. Stream discharges of 1–10 ft^3/s and dilution ratios of liquid formulation and water are given in Table 3.3 for a drip head that dispenses 200 mL/min.

3.2.2 Treatment techniques and equipment

Stream treatments generally require drip stations spaced along the stream at intervals of approximately one mile so a constant amount of rotenone can be added to maintain the desired concentration. A cage containing fish is placed in the stream upstream of

Figure 3.5 Drip barrel system, used by Utah Division of Wildlife Resources, being filled with rotenone and water (top). Drip head dispensing chemical to the stream (bottom).

each drip station (except the first drip station in a series) to ensure that effective concentrations of rotenone reach the drip station. Applicators are assigned to maintain the delivery of rotenone at the drip stations at the desired concentration by monitoring and recharging drip containers.

Additional crew members are assigned to walk each stream section and add additional rotenone to seeps and springs tributary to the stream. This can be accomplished with a rotenone sand mixture (see Appendix O). The sand mixture consists of powdered rotenone, sand, and gelatin. This mixture is effective for treating seeps and springs because it maintains piscicide concentrations in the water longer than applications of liquid. Powder without the sand-gelatin mixture does not reach the oxygenated-water source.

Crews should be given carefully measured amounts of liquid formulation diluted 1:10 with water. The total rotenone used in the treatment should not exceed the maximum dosage on the label. This is also critical when neutralization is required downstream.

3.2.3 Use of biological monitoring tests to determine efficacy

Target fish should be held in live-cages immediately upstream from each drip station and above the neutralization station. These fish should be monitored to determine the time between the start of the upstream drip and when the fish begin to show signs of stress and are killed. Death of the fish will provide the assurance that the treatment has been successful. See Sections 3.5.3 and 3.5.4 for testing efficacy of the neutralization activities.

3.3 APPLICATOR SAFETY

A safe work environment for personnel involved in the application of rotenone is a critical part of fishery management projects. The hazards associated with rotenone use can be mitigated if applicators are properly trained and utilize the proper equipment.

Table 3.3. Application chart for liquid rotenone at concentrations of 1.0 and 2.0 ppm for 4, 6, and 8 h. For this table the drip head dispenses at a rate of 200 mL/min or 12.68 gal in 4 h, 19.02 gal in 6 h and 25.36 gal in 8 h). Values in gallons; R = rotenone.

	4 h				6 h				8 h			
	1.0 ppm		2.0 ppm		1.0 ppm		2.0 ppm		1.0 ppm		2.0 ppm	
ft^3/s	R	Water	R	Water	R	Water	R	Water	R	Water	R	Water
1	0.108	12.572	0.216	12.46	0.162	18.858	0.324	18.696	0.216	25.144	0.432	24.928
2	0.216	12.464	0.432	12.25	0.324	18.696	0.648	18.372	0.432	24.928	0.864	24.496
3	0.324	12.356	0.648	12.03	0.486	18.534	0.972	18.048	0.684	24.676	1.368	23.992
4	0.432	12.248	0.864	11.81	0.684	18.336	1.368	17.652	0.864	24.496	1.728	23.632
5	0.54	12.14	1.08	11.6	0.81	18.21	1.62	17.4	1.08	24.28	2.16	23.2
6	0.648	12.032	1.296	11.38	0.972	18.048	1.944	17.076	1.296	24.064	2.592	22.768
7	0.756	11.924	1.512	11.17	1.134	17.886	2.268	16.752	1.512	23.848	3.024	22.336
8	0.864	11.816	1.728	10.25	1.296	17.724	2.592	16.428	1.728	23.632	3.456	21.904
9	0.972	11.708	1.944	10.74	1.458	17.562	2.916	16.104	1.944	23.416	3.888	21.472
10	1.08	11.6	2.16	10.52	1.26	17.76	2.52	16.5	2.16	23.2	4.32	21.04

3.3.1 Applicator hazards

Rotenone products used in fisheries management have been classified by the USEPA as Category 1 materials, which are in the "extremely toxic" range for acute (short-term) toxicity. Inhalation, dermal, and ocular exposures are the three most common routes of applicator exposure. These types of exposures are significantly mitigated by the use of proper handling procedures and protective equipment such as air-purifying respirators, protective clothing (coveralls, gloves), and eye protection (splash goggles or face shields). Specific information on proper handling procedures and protective equipment are found in the sections specific to powder and liquid formulations.

The USEPA is the primary regulatory authority for pesticide applicator safety in the United States. The Pest Management Regulatory Agency (PMRA) is the appropriate authority in Canada. Pesticide applicators are legally required to follow the applicator safety precautions indicated on USEPA-approved or PMRA-approved product labeling. This labeling includes the safety precautions and protective equipment that must be employed during the use of the pesticide product. In addition to the federal safety requirements found on the product labeling, some states may require additional applicator safety procedures and equipment. These state-mandated requirements are usually more stringent than the federal standards. In general, a policy of following the most stringent requirement, whether federal or state-mandated, should be adopted.

3.3.1.1 Applicator supervision

The use of formulated rotenone products must be supervised on-site by at least one person who has federal or state certification as a pesticide applicator. These project supervisors must have, at the minimum, the authority to start and stop the rotenone application and be well versed in all federal or state regulatory requirements regarding the safe and legal use of the rotenone product and applicator safety.

3.3.1.2 Pesticide safety training

All personnel involved with the rotenone application must receive safety training specific to the formulated rotenone products that will be used and specific to the federal or state regulatory requirements. At the minimum, this safety training includes information on the following: (1) how to read and understand the pesticide product label; (2) the acute and chronic applicator exposure hazards; (3) routes of pesticide exposure; (4) symptoms of overexposure; (5) how to obtain emergency medical care; (6) decontamination procedures; (7) how to use the required safety equipment; (8) safety requirements and proper procedures for pesticide handling, transportation, storage, and disposal; and (9) environmental hazards. Records of this training must be maintained according to the applicable federal or state regulatory requirements.

3.3.1.3 Washing facilities

Applicators will be provided with materials listed below for emergency washing. The amount of wash water should be sufficient for the emergency washing of at least one person. A 5-gal plastic jug (narrow mouth, sealed with lid), a small eye wash bottle (one pint or quart), liquid soap, disposable towels, and extra coveralls should be sufficient. An unlimited water source (water truck or faucet), liquid soap, disposable towels, extra coveralls, and large emergency eyewash station should be available at the central staging area for more complete washing. Moist towelettes (i.e., baby wipes) can be kept on boats so applicators can periodically clean facemasks and quickly remove any dermal contamination, especially when using powdered rotenone formulations.

3.3.1.4 Emergency medical care

For large-scale projects, particularly those that require the coordinated activities of numerous personnel, loading, and application equipment, it is highly recommended that an emergency medical technician (EMT) or other medical professional be on-site during the project. The EMT is to respond to any medical emergencies including those related to formulated rotenone exposure. Additionally, the name, address, and telephone number of a local physician, clinic, or hospital that can provide care in the event of a pesticide emergency should be posted in a prominent location at the project headquarters and at the mixing and loading site.

3.3.1.5 Emergencies or illnesses

Personnel that suspect that they have been made ill by the rotenone must immediately report to the on-site EMT, if available. The certified project supervisor should be immediately notified of the situation. If necessary, the person will remove his or her work clothing, shower, and dress in clean clothing. In the event of rotenone exposure to the eyes, the person's eyes should be immediately flushed with large amounts of water for at least 15 min. The person is attended by the onsite EMT and, if necessary, is taken to the appropriate medical care facility. Copies of the product labels and material safety data sheets for the formulated rotenone products will be provided to both the on-site EMT and to any other medical professionals as needed.

3.3.2 Powder formulation hazards

Powder formulations are generally applied as a rotenone and water slurry. The greatest potential for applicator exposure (dermal, ocular, and inhalation) occurs when the powder compound can become airborne as it is removed from its original container and is mixed with water.

The generation of airborne rotenone powder is significantly reduced by using proper mixing and loading procedures. Powdered rotenone can be removed from original containers by several methods, including hand pouring. However, the size of the containers frequently makes hand pouring impractical or impossible. Removing the powder by vacuum hose is preferable because it reduces airborne powder generation. A Plexiglas or other heavy plastic shield can be put over the open container during the slurry mixing process. A small opening in the shield allows free movement of the vacuum nozzle. Vacuum nozzles should be of durable construction (polyvinyl chloride [PVC]) and should be long enough to eliminate the need for applicators to lean over the container and increase the potential for airborne powder exposure as the container begins to empty.

To reduce rotenone powder contamination of the boat docking area, the rotenone containers are only opened while on the boat. Powdered rotenone containers have an interior plastic liner that can be emptied and rinsed while on the boat. These emptied and rinsed liners should be placed in a master trash bag that should be tied off before the boat returns to the docking area. Product labels require disposal of the containers by incineration or in a landfill.

3.3.3 Liquid formulation hazards

Liquid formulations are generally applied as 10% dilutions with water. The greatest potential for applicator exposure (dermal, ocular, and inhalation) occurs when the concentrated formulation is removed from its original container and mixed with water.

Hazard to the applicator from concentrated rotenone liquid is significantly reduced by using the proper mixing and loading procedures. Liquid formulations can be removed from original containers by hand pouring or pumping. However, hand pouring greatly increases applicator hazard. Additionally, the size of the containers frequently makes hand pouring impractical or impossible. In these cases, pumping the concentrated liquid from its container into a spray tank on the application boat, truck, or all-terrain vehicle is preferable.

3.3.4 Safety equipment

Labeling requires that all personnel who handle opened containers of rotenone or participate in the application of rotenone must use the following protective equipment: (a) coveralls (disposable Tyvek® or reusable cotton); (b) eye protection (splash goggles or full-face protection provided by the full-face respirators indicated below); and (c) nitrile gloves.

Additionally, personnel working with (mixing, loading, or applying) undiluted liquid formulations must use air-purifying respirators with organic vapor-removing cartridges with prefilters approved for pesticides (standard cartridge MSHA/NIOSH prefix TC-23C). Personnel working with (mixing, loading, or applying) powder formulations must use respirators with dual cartridges fitted with dust-filtering cartridges (MSHA/NIOSH prefix TC-21C) and organic vapor–removing cartridges (MSHA/NIOSH prefix TC-23C). Because of the hazard of airborne rotenone, personnel applying powdered rotenone formulations should wear full-face respirators. Personnel required to wear respirators must be provided with respiratory protection training that includes instruction on how to properly fit and test a respirator.

All equipment must be clean and in good repair at the start of each workday. Personnel handling rotenone should be issued two sets of coveralls and gloves. Ripped clothing or damaged equipment must be replaced as soon as possible. Extra sets of coveralls, gloves, and eye protection (equal to approximately 50% of the total number of rotenone handlers) must be available in the event of equipment damage. Respirator cartridges should be changed as needed. Extra respirators and cartridges must be available at the project site.

3.3.5 Monitoring of applicator inhalation hazard

Inhalation of the vapors from liquid rotenone or of dust from powdered rotenone is the most common route of applicator exposure. While the proper use of air-purifying respirators will adequately mitigate the inhalation hazard, applicator monitoring provides an accurate assessment of the individual application inhalation hazard.

For powdered rotenone applications, air samples can be collected via air pumps from the breathing zones of personnel that mix, load, or apply the rotenone. Air samples are collected at a calibrated rate during a predetermined time (i.e., 8 h). An in-line sampling tube with a Teflon media (1.0-mm pore size) is used (see Section 3.4.3.3.1).

For liquid rotenone, air sampling can measure for rotenone and the volatile and semivolatile nonrotenoid organic constituents of the formulated products (i.e., methyl naphthalene, naphthalene, trichloroethylene, and xylene). Rotenone, naphthalene, trichloroethylene, and xylene samples can be collected using sampling pumps similar to those employed for powdered rotenone monitoring. Xylene and trichloroethylene samples can be collected using passive sampling badges (charcoal media) that are exposed for eight hours (see Section 3.4.3.3.1).

Once collected, sampling media must be stored and analyzed according to established protocols. The results of analysis will provide an accurate assessment of the inhalation hazard from the chemical compounds of concern. It should be noted, however, that the calculated exposure is for persons who are not employing respiratory protective equipment. The use of air-purifying respirators will provide a 10-fold to 50-fold protection factor.

3.4 MONITORING PROCEDURES

Fish survival and four types of chemical monitoring procedures are used in conjunction with rotenone applications. These include: (1) determining the composition of rotenone formulations; (2) surface and ground water quality monitoring; (3) sediment monitoring; and (4) monitoring air quality following rotenone applications. These activities establish the effectiveness of the treatment; the degradation, neutralization, and dispersion of chemicals; and applicator exposure. The location and size of the treatment and the surrounding environment dictate the complexity of the monitoring program.

The entire affected area should be outlined on USGS topographic quadrangle maps (1:24,000 scale) before designing a chemical monitoring plan. The location of treatment sites, potable water supply intakes, other water diversions, groundwater supplies, and other areas of environmental significance including areas of high public use (i.e., parks, etc.) must be indicated. The treatment area is subdivided into zones of treatment dates and times to assist with managing and monitoring activities. The monitoring program has quality assurance and quality control elements to ensure reasonable accuracy and precision for the chemical data. Control limits for accuracy and precision are decided before the study begins.

3.4.1 Fish survival

Caged fish can be placed at various locations and depths in a lake or stream to monitor treatment efficacy. Caged fish are then monitored from the time the treatment begins until the fish die. This procedure will indicate the time required to kill the fish and the percentage (effectiveness) of the kill. For sampling projects, it is necessary to collect and mark or stock a known number of marked fish into the sample site before the kill to estimate the percentage of recovered fish within the sample site. Following the treatment, caged fish are placed in the water to determine how long the chemical remains toxic. This may take from 1 day to 4 weeks depending on water temperature, pH, and alkalinity. When fish live for 48 h in the cages it is generally considered safe to restock the water body.

3.4.2 Chemical composition of liquid rotenone formulations

Each lot of formulated rotenone is analyzed for (1) percent active ingredients, (2) concentrations of inert solvents, carriers, and emulsifiers, and (3) presence of contaminants. These constituents are called inert because they contribute no toxicity to the formulation. Normally, the manufacturer will certify the percent active ingredient at time of production. It is usually not necessary to reanalyze new products, but this should be done if material is older than 1 year. The rotenone content is measured using methods of Dawson et al. (1983) and the synergist piperonyl butoxide (PBO) is analyzed by USEPA method 8270B (USEPA 1994a). The inert ingredients and contaminant concentrations are quantified using methods for volatile organic compounds (VOC; USEPA methods 8260B or 624 [USEPA 1994b; USEPA 1984a]) and semivolatile organic compounds (semiVOC; methods 8270B [USEPA 1994a] or 625 [USEPA 1984b]).

Chemical analyses of different lots of Nusyn-Noxfish® have found inert ingredients that include trichloroethylene (TCE) (10–1,200 ppm), naphthalene (60,000–120,000 ppm), 2-methylnaphthlene (100,000–120,000 ppm), toluene (200–560 ppm), ethylbenzene (300–900 ppm), and xylene (80–5,400 ppm) (CDFG 1994; Siepmann and Finlayson 1999). The inert ingredients make up approximately 85% to 90% of the liquid formulations. The types and percentages of various inert ingredients in the formulations are changed by the manufacturer. Additionally, the formulation of Nusyn-Noxfish® contains the synergist PBO at 2.5% (25,000 mg/L). The results of these analyses provide useful information on the amount of chemical required for an effective treatment, the types or frequency of water, sediment, and air monitoring samples, and the potential impacts on water and air quality.

It may be necessary to monitor rotenone concentrations throughout the treatment area and to monitor the degradation of rotenone over time. This information will assist in determining the effectiveness of the treatment and when public contact and restocking with fish may occur. It may be desirable to monitor rotenone residues downstream from the addition of a neutralizing agent (i.e., potassium permanganate) to confirm successful neutralization. Monitoring the VOC and semiVOC concentrations in the treatment area can provide useful information related to water quality and public safety.

It takes several hours to several days for complete mixing of rotenone in large impoundments and lakes, depending on depth and the type of formulation applied. Complete mixing occurs faster in shallower water bodies using liquid rotenone than in deeper water bodies using powder. Average (throughout water column) rotenone concentrations will approach expected levels after complete mixing. Liquid formulation disperses quickly in streams, usually within 100 yards, depending on flow and turbulence.

The use of powdered and liquid rotenone formulations will require different monitoring procedures because of different chemical constituents. Powdered rotenone contains only ground plant root and therefore analyses for rotenone and its primary degradation product rotenolone are sufficient. The two common commercial liquid formulations Nusyn-Noxfish® and Noxfish® contain inert emulsifiers, solvents, and carriers that are important in ensuring the solubility and dispersion of rotenone in water.

A number of rotenone applications have been monitored over the last 10 years (CDFG 1994; Siepmann and Finlayson 1999; Finlayson and Harrington, unpublished data, 1991). Rotenone, rotenolone, naphthalene, and methylnaphthalene have been found in water and sediment. Xylene, TCE, and PBO have been found in water. The type of rotenone formulation used will dictate whether water or sediment are monitored and for which chemicals (Table 3.4). For a more complete discussion of the issues associated with these residues see Section 5.

3.4.3 Environmental media

Chemical analysis of water and sediment samples for rotenone content may be required, depending on the location of the treatment and use of the water. Some states require analysis of water samples periodically following the treatment until all rotenone, rotenone products, and other formulation products are no longer detectable. In such cases a monitoring plan must be developed before the treatment (see Section 5). Most

Table 3.4. Matrix for selection of analytical methods based on rotenone formulation and environmental media. VOC = volatile organic compound.

Media	Pro-Noxfish®	Nusyn-Noxfish®	Noxfish®
		Water	
Dawson et al. (1983)	rotenone and rotenolone	rotenone and rotenolone	rotenone and rotenolone
USEPA meth 8310		semi VOC	semi VOC
USEPA meth 502.2		VOC	VOC
USEPA meth 8270B		piperonyl butoxide	
		Sediment	
Dawson et al. (1983)	rotenone and rotenolone	rotenone and rotenolone	rotenone and rotenolone
USEPA meth 8270B		semi VOC	semi VOC
USEPA meth 8260A		VOC	VOC
USEPA meth 8270B		piperonyl butoxide	

agencies contract with independent laboratories for sample collection and analysis. The CDFG (1994) and Siepmann and Finlayson (1999) provide excellent discussions of water monitoring techniques.

3.4.3.1 Ground and surface water

3.4.3.1.1 Water sampling frequency—Water and sediment in surface and groundwater sites must be sampled to establish background levels of rotenone, rotenolone, and other organic chemicals that may be in the formulations. Surface water sites are sampled during (flowing water) or immediately following (standing water) the rotenone application at established intervals. Sampling normally continues until rotenone has dissipated from the treatment area. Groundwater sites are typically sampled 1 week, 3 weeks, 3 months, 6 months, and 12 months following treatment.

3.4.3.1.2 Water collection—Samples for rotenone and rotenolone concentrations in water are taken in chemically clean, 500-mL amber glass bottles. Samples for VOC concentrations in water are collected in chemically clean, 40-mL vials with septa caps. Samples for semiVOC and PBO concentrations in water are taken in clean 1-L amber glass bottles. All of the glass containers have Teflon-lined caps.

Surface samples are taken by submerging the capped bottles a few inches below the surface, uncapping the bottle and allowing it to fill, and recapping the bottle below the surface. Care should be taken to exclude air space in the bottles. Subsurface samples at any depth are collected using a Kemmerer bottle. Groundwater from established wells should be sampled as close to the wellhead as possible by sampling from a Schrader valve or faucet before the storage tank using standard procedures (Sava 1986). Well pumps are turned on for a minimum of 15 min to purge standing water in the well casing.

Consider collecting samples in replicate and analyzing one of the two replicates. The replicates serve as insurance against breakage during transit or an analytical anomaly. Because temperature and pH affect the degradation rate of rotenone, it is advisable to record these water quality parameters. The samples should be placed on ice immediately after collection, transported to a refrigerator at the laboratory, and stored at 4°C until analyzed. The allowable holding times for samples vary from 7 to 14 d.

3.4.3.1.3 Water sample analyses—Rotenone and rotenolone are measured by the methods of Dawson et al. (1983). VOC are detected by USEPA method 502.2 (USEPA 1989a), semiVOC are measured by USEPA method 8310 (USEPA 1986), and PBO is analyzed by USEPA method 8270B (USEPA 1994a).

3.4.3.2 Sediment

3.4.3.2.1 Sediment sampling frequency—All of the sites must be sampled to establish background levels of rotenone, rotenolone, and other organic chemicals that may be in the formulations. Sites are initially sampled during (flowing water) or immediately following (standing water) the rotenone application at previously established intervals until rotenone has dissipated from the treatment area. Sediment is tested less often than water, and concentrations seem to lag about one week behind water concentrations (Siepmann and Finlayson 1999; Finlayson and Harrington, unpublished data, 1991). Sediment sampling should be no more frequent than once a week.

3.4.3.2.2 Sediment collection—Collect samples in chemically clean, 500-mL amber glass jars. Collect separate samples for rotenone and rotenolone, VOC, semiVOC, and PBO analyses. Only use glass sample jars with Teflon-lined lids. Collect at least 100 mL of sediment by scraping sediment off the bottom of the lake or stream. Fill the remainder of the jar with overlying water. Exclude air space in the jars and lids. Deeper samples can be collected using a sediment core sampler. Consider collecting samples in replicate. Place samples on ice immediately after collection, transport to a refrigerator at the laboratory, and maintain at a temperature of 4°C until analyzed. The allowable holding times for samples vary from 7 to 14 d.

3.4.3.2.3 Sediment analyses—Rotenone and rotenolone are measured using the method of Dawson et al. (1983); VOC are detected by USEPA method 8260B (USEPA 1994b), semiVOC are detected by USEPA method 8270B (USEPA 1994a), and PBO is measured by USEPA method 8270B (USEPA 1994a).

3.4.3.3 Air

Air monitoring studies following rotenone applications are completed only if there is a demand. Monitoring of air may be done to document exposure levels of certain compounds to rotenone applicators or to the public in nearby areas. Some of the compounds in Pro-Noxfish®, Nusyn-Noxfish® and Noxfish® have permissible exposure limits (PEL) established by federal and state regulation. The current Occupational Safety and Health Administration (OSHA) PEL for rotenone (OSHA 1978) and TCE (OSHA 1988) are 5 mg/m^3 and 100 mg/L, respectively, as time-weighted averages for an 8-h period.

3.4.3.3.1 Air sample collection—Pumps draw air through an activated charcoal tube for lighter-weight hydrocarbons (TCE and xylene), a particulate Teflon filter for rotenone, and XAD-2 tubes for heavier molecular weight hydrocar-

Table 3.5. Matrix for selection of sampling device for air based on the rotenone formulation.

Media	Pro-Noxfish®	Nusyn-Noxfish®	Noxfish®
		Air	
Teflon filter	rotenone and rotenolone	rotenone and rotenolone	rotenone and rotenolone
XAD-2 tube		heavier hydrocarbons[1]	heavier hydrocarbons[1]
Charcoal tube		lighter hydrocarbons[2]	lighter hydrocarbons[2]

[1] For heavier molecular weight hydrocarbons; roughly equivalent to semi-VOC (volatile organic compound).

[2] For lighter molecular weight hydrocarbons; roughly equivalent to VOC.

bons (naphthalene and methylnaphthalene) during a 12–24-h period. The tubes and filters are collected and placed on dry ice until analyzed. The rotenone formulation will dictate the type of tubes to use (Table 3.5).

3.4.3.3.2 Air analyses—A desorbing solvent is used to remove the compounds from the tubes and filters. Benzyl alcohol is used for the charcoal tubes, acetonitrile is used for Teflon particulate filters, and hexane is used for the XAD-2 tubes. The compounds from the XAD-2 tubes are analyzed by gas chromatography/mass spectrometry (GC/MS) with a detection limit of 1 mg/m^3 naphthalene; the compounds for the Teflon particulate filters are analyzed by High Performance Liquid Chromatograph (HPLC) with a detection limit of 0.003 mg/m^3 rotenone; and the compounds from charcoal tubes are analyzed by GC/MS with detection limits of 3 mg/m^3 TCE (CARB 1997).

3.5 NEUTRALIZATION PROCEDURES FOR CHEMICAL TREATMENT PROJECTS

Rotenone usually degrades naturally within one to four weeks depending on pH, alkalinity, temperature, and dilution with the untreated water (Schnick 1974). The need for neutralization of discharge water depends on the use of the water, timing, and other considerations. If the discharge cannot be stopped without impacts to downstream users, neutralization is necessary. The chemical most commonly used to neutralize rotenone formulations is potassium permanganate ($KMnO_4$). Most other agents that can neutralize rotenone have deleterious attributes that make them far less useful for this purpose. For example, the label allows chlorine for neutralization, but it may not meet state water quality standards.

3.5.1 Neutralization problems

Considerable research and practical experience in the application of rotenone and its efficacy in eradicating a fish community has been accomplished, but there have been few controlled experiments on neutralizing the active ingredients of rotenone to avoid killing fish and other organisms outside the target area or to permit reintroduction of aquatic organisms following treatments. Attempts to neutralize the piscicide have been notoriously unpredictable and occasionally have resulted in serious and highly publicized failures. A major problem has been the lack of understanding of the factors influencing the efficacy of the neutralizing agent in the highly variable chemistry of natural waters. Difficulties in making accurate measurements of the rotenone concentrations to be neutralized have also lead to problems. Careful attention to application rates and limnological parameters can achieve a high level of success in neutralization projects.

3.5.2 Neutralization with potassium permanganate

Potassium permanganate is a strong oxidizer that breaks down into potassium, manganese, and water. All are common in nature and have no deleterious environmental effects at the concentrations normally associated with the neutralizing processes.

The use of $KMnO_4$ requires precautions to ensure the safety of applicators and to avoid spontaneous combustion. The chemical is caustic to the mucous membranes of the nose and throat and causes brown stains on the skin and clothing on contact when dissolved in water. Potassium permanganate is dusty and should not be handled without protective clothing and breathing apparatus.

Potassium permanganate is usually packaged in 110-lb drums in a fine to coarse granular form. A 5% aqueous solution is also available. The coarse granular form is less expensive but more difficult to dissolve than fine granules. The 5% aqueous solution of $KMnO_4$ is available from one known source: Zoom Environmental Service Inc. (Sterling Heights, Michigan) at roughly US$3.15/gal in 55-gal barrels. The dry forms generally cost between $1.50 and $2.50/lb (1999 cost) of 99% pure product and are available from several suppliers of industrial or laboratory chemicals. The cost to treat one acre-foot of water at one ppm with a granular form would range from $0.20 to $0.34, compared to $1.04 for the 5% aqueous solution (1999 cost).

Dissolve the granular product in water, then apply the solution to the stream or apply the dry granular form directly to the stream. If the product is dissolved in water before application, granulation is not critical, but if applied directly to the stream as a dry preparation, the fine free-flowing product produced by Carus Chemical Company (Ottawa, IL 61350) is preferable because it feeds freely from the dispensing hopper.

The chemical must be kept away from organic materials such as gasoline, oils, alcohols, or any other oxidizable material. It also reacts with many metals when dissolved, so it is packaged in steel- or nickel-coated containers. The dry material is inert, but becomes active once dissolved in water. If the chemical comes in contact with the eyes or skin, the area should be flushed with copious amounts of water.

Potassium permanganate is soluble in water with saturation achieved in distilled water with 0.36 lb/gal at 10°C and 0.54 lb/gal at 20°C. Activity is influenced by temperature—neutralization is slowed at low temperatures and accelerated at high temperatures. Most applicators work in moderate temperature ranges between 10°C and 25°C and don't consider temperature in estimating neutralization times, but applicators should take into account that chemical reactions slow by 50% for each 10°C reduction in temperature and double for each 10°C increase. This factor is especially important in determining the stream reach and neutralization lag times at very low temperatures.

$KMnO_4$ is toxic to fish at relatively low concentrations under some circumstances and is much more toxic in alkaline water than soft water (Marking and Bills 1975) (Figure 3.6). $KMnO_4$ breaks down in the natural environment quite rapidly and is a much-preferred alternative to the dispersion of a toxic plume of rotenone many miles downstream of the target area as opposed to a short plume of toxic $KMnO_4$ immediately below the target zone. If $KMnO_4$ concentrations are in balance with rotenone concentrations then toxic levels of $KMnO_4$ should be quickly reduced through the oxidation of organic components and rotenone in the water.

3.5.3 Application rates

Incidents of accidental release of rotenone into nontarget areas have been attributed to the erroneous rule of thumb that rotenone is neutralized with $KMnO_4$ at a 1:1 ratio. This rule is approximately true for distilled water, but several components of most natural waters alter the relationship. Engstrom-Heg (1971, 1972) conducted controlled experiments dem-

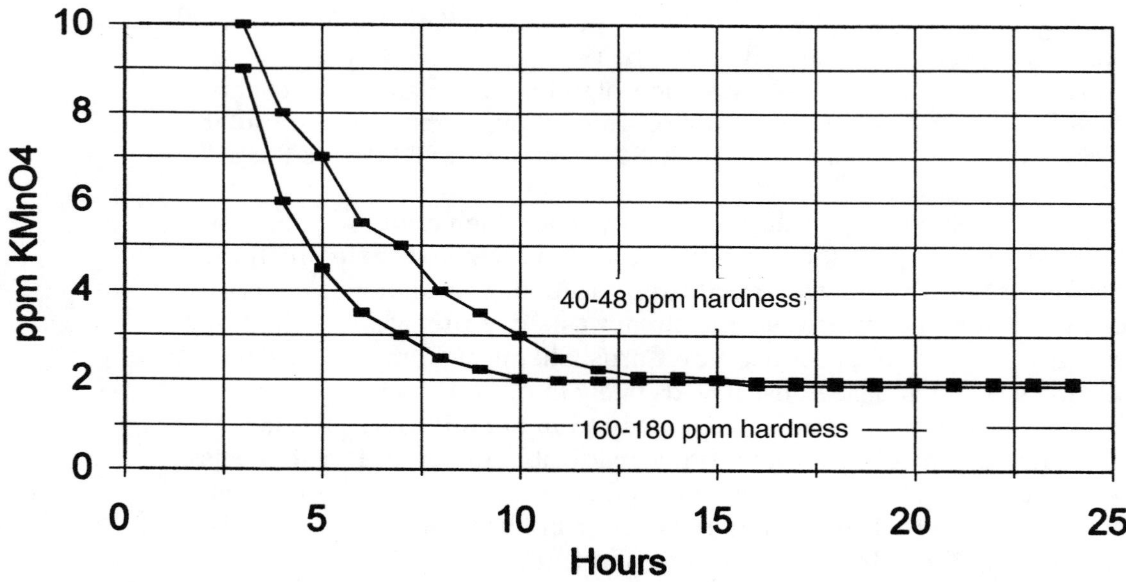

Figure 3.6 Toxicity of $KMnO_4$ to rainbow trout at 12°C at 40-48 and 160-180 mg/L total hardness (adapted from Marking and Bills 1975).

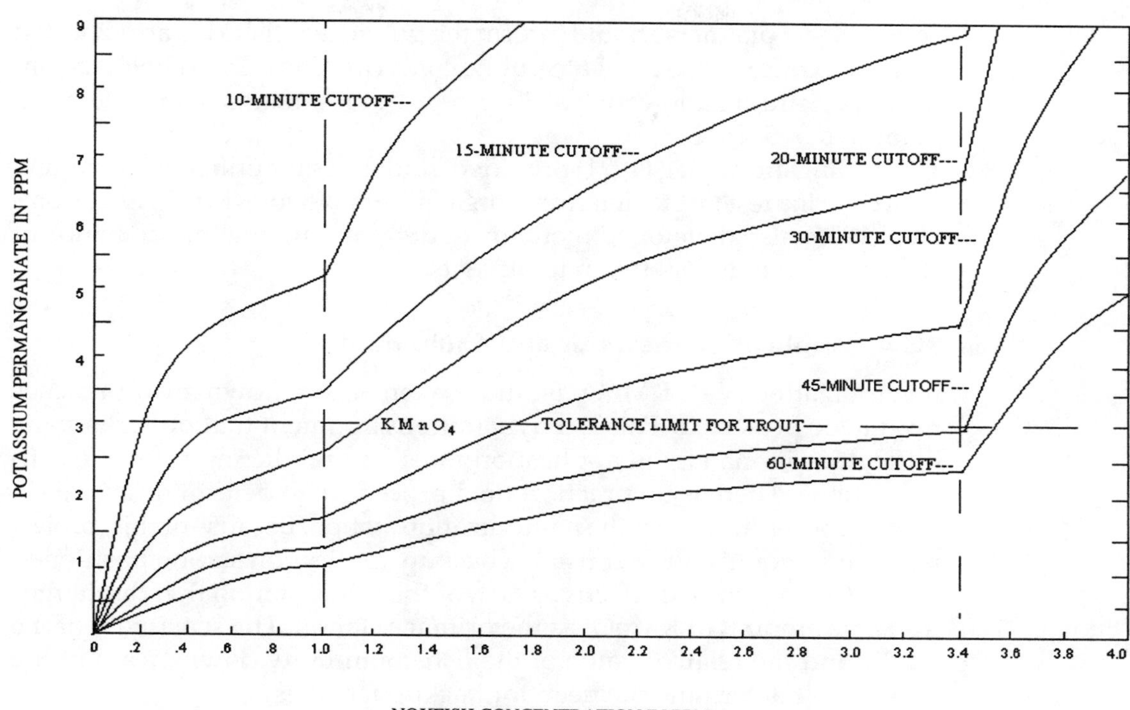

Figure 3.7 Relationship between Noxfish® concentration (5% rotenone), $KMnO_4$ concentration, and contact time required for neutralization (from Engstrom-Heg 1972).

onstrating that dissolved electrolytes and suspended organic matter have a major influence on the amount of $KMnO_4$ required to neutralize a given concentration of rotenone. He developed time-lapse curvilinear relationships for neutralizing 5% Noxfish® concentrations over a range of $KMnO_4$ concentrations in distilled water (Figure 3.7).

Engstrom-Heg (1971, 1972) expanded those relationships for natural waters by developing a formula to account for organic demand and total hardness generally encountered in natural systems. He also developed a convenient field method of determining organic demand for the treated waters. Engstrom-Heg (1972) showed that the amount of $KMnO_4$ required to neutralize a given concentration of rotenone (Figure 3.7) needs to be multiplied by the product of the following formula to account for organic and electrolyte demands:

$$\text{Multiplier} = 1 + 0.002\,(\text{total alkalinity [as ppm } CaCO_3] - 20) + 0.5\,(\text{organic demand [as ppm]})$$

A recent field application in Utah on a stream with a 3 ppm organic demand and 238 ppm total alkalinity required an application of 3.95 ppm $KMnO_4$ to neutralize 2 ppm rotenone formulation in 60 min (D. Archer, UDWR, personal communication, 1999). The CDFG (1994) recommends using a ratio from 2:1 to 4:1 ($KMnO_4$: formulated rotenone) for neutralization. They found that 3 ppm was required to neutralize 1 ppm Nusyn-Noxfish® in a high-elevation stream with low hardness and alkalinity.

Because neutralization of rotenone is not immediate, Horton (1997) recommends considering a neutralization zone below the application point. He quotes Colorado and Utah project findings in which the zone of neutralization was the distance that water can be expected to travel in 20 min. Potential fish mortalities can be expected in this zone. Project planners should inform the public that fish kills are expected in this mixing zone to alleviate public concerns. The CDFG (1994) recommends a neutralization zone of 30 min in alpine streams with water temperatures of 5–15°C.

Engstrom-Heg (1971) provided guidelines for utilizing swimming pool chlorine test kits to determine organic demand for $KMnO_4$. Engstrom-Heg (1976) also developed methods of determining organic demands of stream sediments for stream treatments.

3.5.4 Application methods and equipment

Neutralization with $KMnO_4$ is undertaken where downstream reaches need to be protected. This is a dynamic environment that presents some difficulties in managing application rates and predicting rotenone concentrations. There is no practical field procedure to determine actual rotenone concentrations at the neutralization station. Because of this, project managers should consider using a backup application station positioned downstream from the effective reach of the initial station that discharges the same amount of $KMnO_4$ as the primary station. The sensitivity of the public and the relative value of the fish community downstream of the project will determine the need for backup facilities.

Many state agencies begin neutralization procedures at the onset of rotenone injection to assure that no rotenone passes the neutralization station and to reduce oxygen demand of the streambed immediately downstream from the neutralization station. Other state agencies determine the time when rotenone is expected to arrive at the neutralization stations and begin application of $KMnO_4$ at that time. A series of caged live fish are located (1) above the primary neutralization station to indicate that rotenone has reached that point, (2) just above the second neutralization station to determine the adequacy of the first application of $KMnO_4$, and (3) below the reach of the second neutralization station to provide assurance that total neutralization was achieved.

Continual maintenance of the neutralization stations is required throughout the treatment project. A supervisor can provide assistance to stations as needed for breaks, errands, and loading and monitoring live fish cages. Each worker maintains a log of activities, application rates, and observations to assure that prescribed procedures are followed and for future reference.

3.5.5 Equipment

Many innovative devices have been developed to dispense $KMnO_4$ in liquid form. Most use a reservoir with some form of metering device to dispense the reagent at a constant measurable rate (for details see Appendix P).

3.6 FISH COLLECTION AND DISPOSAL

Most chemical treatments in standing water are completed in late summer or fall when dead fish are not a health hazard or environmental threat. In warmer latitudes, sampling projects are conducted from early spring to late fall. Stream treatments (including stream sampling) generally occur in the late summer and fall when flows are low and dead fish are not a significant problem. The unwritten position of the UDWR is that dead fish will not be collected following a treatment project (Thompson, personal communication, 1999). The Michigan Department of Natural Resources policy (MDNR 1993) states that recovery of dead fish will not normally be attempted by the Fisheries Division. However, certain circumstances justify Division assistance in the collection and removal of dead fish, particularly if there is little cost to the Division or if a recovery program is organized by private landowners. In such cases cooperation is encouraged. Horton (1997) states the Idaho Department of Fish and Game policy is to leave dead fish in the treated water. The policy of CDFG (1994) is to remove fish from the treatment area when dead fish may become a public nuisance or when a request is made by a public agency. All states contacted reported that dead fish are recovered to avoid serious public controversy in sensitive situations.

Fish collection during and after standing water treatments is accomplished with multiple boats and dip nets. Following treatment, crews are organized to patrol the shore and collect dead fish. In fish sampling situations, fish are generally picked up for 2–5 d, dependent on water temperature. Fish collections from stream treatments are most effective using block nets. Dead fish must be continually removed from block nets or they will eventually collapse them, allowing fish to go downstream. Good management of this process will reduce public relations problems and not give the appearance of a fish kill beyond the intended treatment zone. Fish are then transported to a disposal site. If fish collection is conducted it should be well planned and executed. Arrangements must be made in advance of the treatment to locate a suitable permitted disposal location. Dead fish are not to be offered or provided for human consumption because the U.S. Food and Drug Administration has not approved this practice.

Appendix H

Noxfish®, Nusyn-Noxfish®, and Pro-Noxfish® Labels (AgrEvo) and Prenfish®, Synpren-Fish®, and Rotenone Powder Labels (Prentiss)

SPECIMEN LABEL

CODE B4674

RESTRICTED USE PESTICIDE
Due to Aquatic Toxicity and Acute Inhalation

For retail sale to, and use only by, Certified Applicators or persons under their direct supervision and only for those uses covered by the Certified Applicator's certification.

NOXFISH®

Fish Toxicant
Liquid-Emulsifiable

* ***FOR CONTROL OF FISH IN LAKES, PONDS, RESERVOIRS AND STREAMS***

ACTIVE INGREDIENTS:

Rotenone	5.0% w/w
Other Associated Resins	5.0%
OTHER INGREDIENTS:*	90.0%
	100.0% w/w

*This product contains aromatic hydrocarbons.
®Noxfish is a registered trademark of AgrEvo Environmental Health, Inc.

EPA REG. NO. 432-172 **EPA EST. NO.**

KEEP OUT OF REACH OF CHILDREN

 **DANGER POISONOUS** 

STATEMENT OF PRACTICAL TREATMENT

IF INHALED: Remove victim to fresh air. If not breathing, give artificial respiration, preferably mouth-to-mouth. Get medical attention.

IF IN EYES: Hold eyelids open and flush with a steady, gentle stream of water for 15 minutes. Get medical attention.

IF SWALLOWED: Promptly drink a large quantity of milk, egg white, gelatin solution or if these are not available, large quantities of water. Avoid alcohol. Do not induce vomiting. Call a physician or Poison Control Center.

IF ON SKIN: Wash with plenty of soap and water. Get medical attention.

See Below For Additional Precautionary Statements.

NET CONTENTS

172s 021299tk

PRECAUTIONARY STATEMENTS
Hazards to Humans & Domestic Animals

DANGER

Fatal if inhaled. May be fatal if swallowed. Harmful if absorbed through skin. Causes substantial but temporary eye injury. Causes skin irritation. Do not breath spray mist. Do not get in eyes, on skin or on clothing. Wear goggles or safety glasses. When working with undiluted product wear either a respirator with an organic-vapor removing cartridge with a prefilter approved for pesticides (MSHA/NIOSH approval number prefix TC-23C), or a canister approved for pesticides (MSHA/NIOSH approval number prefix 14G) or a NIOSH approved respirator with an organic vapor (OV) catridge or canister with any R>P. or HE prefilter.

Wash thoroughly with soap and water after handling and before eating, drinking or using tobacco. Remove contaminated clothing and wash before reuse.

Environmental Hazards

This pesticide is extremely toxic to fish. Fish kills are expected at recommended rates. Consult your State Fish and Game Agency before applying this product to public waters to determine if a permit is needed for such an application. Do not contaminate untreated water when disposing of equipment washwaters.

Chemical and Physical Hazards

FLAMMABLE: KEEP AWAY FROM HEAT AND OPEN FLAME. FLASH POINT MINIMUM 45°F (7°C).

DIRECTIONS FOR USE

It is a violation of Federal law to use this product in a manner inconsistent with its labeling.

STORAGE AND DISPOSAL

Do not contaminate water, food or feed by storage or disposal.
Storage: Store only in original containers, in a dry place inaccessible to children and pets. Noxfish will not solidify nor show any separation at temperatures down to 40°F and is stable for a minimum of one year when stored in sealed drums at 70°F.
Pesticide Disposal: Pesticide wastes are acutely hazardous. Improper disposal of excess pesticide, spray mixture, or rinsate is a violation of federal law. If these wastes cannot be disposed of by use according to label instructions, contact your state pesticide or Environmental Control Agency or the Hazardous Waste representative at the nearest EPA Regional Office for guidance.
Container Disposal: Triple rinse (or equivalent). Then offer for recycling or reconditioning or puncture and dispose of in a sanitary landfill or by other procedures approved by state and local authorities.

Noxfish is registered for use by or under permit from, and after consultation with State and Federal Fish and Wildlife Agencies.

General Information
Noxfish is a specially formulated product containing rotenone to be used in fisheries management for the eradication of fish from lakes, ponds, reservoirs and streams.

Since such factors as pH, temperature, depth and turbidity will change effectiveness, use this product only at locations, rates, and times authorized and approved by appropriate state and federal fish and wildlife agencies. Noxfish is registered for use by or under permit from, and after consultation with State and Federal Fish and Wildlife Agencies. Rates must be within the range specified on the label.

Properly dispose of unused product. Do not use dead fish for food or feed.

Do not use water treated with rotenone to irrigate crops or release within ½ mile upstream of a potable water or irrigation water intake in a standing body of water such as a lake, pond or reservoir.

Re-entry Statement: Do not allow swimming in rotenone-treated water until the application has been completed and all pesticide has been thoroughly mixed into the water according to labeling instructions.

For Use in Ponds, Lakes and Reservoirs

The actual application rates and concentrations of rotenone needed to control fish will vary widely, depending on the type of use (e.g. selective treatment, normal pond use, etc.) and the factors listed above. The table below is a general guide for the proper rates and concentrations.

Noxfish disperses readily in water both laterally and vertically, and will penetrate below the thermocline in thermally stratified bodies of water.

Computation of Acre-Feet: An acre-foot is a unit of volume of a body of water having the area of one acre and the depth of one foot. To determine acre feet in a given body of water, make a series of transects across the body of water taking depths with a measured pole or weighted line. Add the soundings and divide by the number made to determine the average depth. Multiply this average depth by the total surface area in order to determine the acre feet to be treated. If the number of surface acres is unknown, contact your local Soil Conservation Service, which can determine this from aerial photographs.

Amount of Noxfish Needed for Specific Uses: To determine the approximate number of gallons of Noxfish (5.0% Rotenone) needed, find your "Type of Use" in the first column of the table below and then divide the corresponding numbers in the fourth column, "Number of Acre-Feet Covered by One Gallon" into the number of acre-feet in your body of water.

Pre-Mixing and Method of Application: Pre-mix with water at a rate of one gallon Noxfish to 10 gallons of water. Uniformly apply over water surface or bubble through underwater lines.

Detoxification: Noxfish treated waters detoxify under natural conditions within one week to one month depending upon temperatures, alkalinity, etc. Rapid detoxification can be accomplished by adding chlorine or potassium permanganate to the water at the same rate as Noxfish in parts per million, plus enough additional to meet the chlorine demand of the untreated water.

Removal of Taste and Odor: Noxfish treated waters do not retain a detectable taste or odor for more than a few days to a maximum of one month. Taste and odor can be removed immediately by treatment with activated charcoal at a rate of 30 ppm for each 1 ppm Noxfish remaining. (Note: As Noxfish detoxifies, less charcoal is required.)

Restocking After Treatment: Wait 2 to 4 weeks after treatment. Place a sample of fish to be stocked in wire cages in the coolest part of the treated waters. If the fish are not killed within 24 hours, the water may be restocked.

Use in Streams Immediately Above Lakes, Ponds and Reservoirs

The purpose of treating streams immediately above lakes, ponds and reservoirs is to improve the effectiveness of lake, pond and reservoir treatments by preventing target fish from moving into the stream corridors, and not to control fish in streams per se. The term "immediately" means the first available site above the lake, pond or reservoir where treatment is practical, while still creating a sufficient barrier to prevent migration of target fish into the stream corridor.

In order to completely clear a fresh water aquatic habitat of target fish, the entire system above or between fish barriers must be treated. See the use directions for streams and rivers on this label for proper application instructions.

General Guide to the Application Rates and Concentrations of Rotenone Needed to Control Fish in Lakes, Ponds and Reservoirs[1]

Type of Use	Parts Per Million		Number of Acre-Feet Covered by One Gallon
	Noxfish	Active Rotenone	
Selective treatment	0.10 to 0.13	0.005 to 0.007	30 to 24
Normal pond use	0.5 to 1.0	0.25 to 0.50	6.0 to 3.0
Remove bullheads or carp	1.0 to 2.0	0.050 to 0.100	3.0 to 1.5
Remove bullheads or cap in rich organic ponds	2.0 to 4.0	0.100 to 0.200	1.5 to 0.75
Preimpoundment treatment above dam	3.0 to 5.0	0.150 to 0.250	1.0 to 0.60

[1] Adapted from Kinney, Edward. 1965. Rotenone in Fish Pond Management. USDI Washington, D.C. Leaflet FL-576.

NOXFISH®

In order to treat a stream immediately above a lake, pond or reservoir, you must: (a) select the concentration of active rotenone, (b) compute the flow rate of the stream, (c) calculate the application rate, (d) select an exposure time, (e) estimate the amount of product needed, (f) follow the method of application. To prevent movement of fish from the pond, lake or reservoir, stream treatment should begin before and continue throughout treatment of the pond, lake or reservoir until mixing has occurred.

1. Concentration of Active Rotenone
Select the concentration of active rotenone based on the type of use from those listed on the table. Example: If you select "normal pond use", you could select a concentration of 0.025 part per million.

2. Computation of Flow Rate for Stream
Select a cross section of the stream where the banks and bottom are relatively smooth and free of obstacles. Divide the surface width into 3 equal sections and determine the water depth and surface velocity at the center of each section. In slowly moving streams, determine the velocity by dropping a float attached to 5 feet of loose monofilament fishing line. Measure the time required for the float to move 5 feet. For fast-moving streams, use a longer distance. Take at least three readings at each point. To calculate the flow rate from the information obtained above, use the following formula:

$$F = \frac{Ws \times D \times L \times C}{T}$$

Where F = flow rate (cubic feet/second), Ws = surface width (feet), D = mean depth (feet), L = mean distance traveled by float (feet), C = constant (0.8 for rough bottoms and 0.9 for smooth bottoms), and T = mean time for float (sec.).

3. Calculation of Application Rate
In order to calculate the application rate (expressed as gallons/second), convert the rate in the table (expressed as gallons/ acre-feet) to gallons per cubic feet and multiply by the flow rate (expressed as cubic feet/second). Depending on the size of the stream and the type of equipment, the rate could be expressed in other units, such as ounces/hour or cc/minute.

The application rate for the stream is calculated as follows:

$$Rs = Rp \times C \times F$$

Where Rs = application rate for stream (gallons/second), Rp = application rate for pond (gallons/acre-feet), C = 1 acre foot/43,560 cubic feet, and F = flow rate of the stream (cubic feet/second).

4. Exposure Time
The exposure time would be the period of time (expressed in hours or minutes) during which Noxfish is applied to the stream in order to prevent target fish from escaping from the pond into the stream corridor.

5. Amount of Product
Calculate the amount of product for a stream by multiplying the application rate for streams by the exposure time.

$$A = Rs \times H$$

Where A = the amount of product for the stream application, Rs = application rate for stream (gallons/second), and H = the exposure time expressed in seconds.

For Use in Streams and Rivers
Only state or federal fish and wildlife personnel or professional fisheries biologists under the authorization of state or federal fish and wildlife agencies are permitted to make applications of Noxfish for control of fish in streams and rivers. Informal consultation with Fish and Wildlife personnel regarding the potential occurrence of endangered species in areas to be treated should take place. Applicators must reference AgrEvo Environmental Health's Noxfish Stream and River Use Monograph before making any application to streams or rivers.

WARRANTY STATEMENT

Our recommendations for use of this product are based upon tests believed to be reliable. The use of this product being beyond the control of the manufacturer, no guarantee, expressed or implied, is made as to the effects of such or the results to be obtained if not used in accordance with directions or established safe practice. The buyer must assume all responsibility, including injury or damage, resulting from its misuse as such, or in combination with other materials.

AgrEvo Environmental Health
95 Chestnut Ridge Road
Montvale, NJ 07645

SPECIMEN LABEL

RESTRICTED USE PESTICIDE
Due to Aquatic Toxicity and Acute Inhalation

For retail sale to, and use only by, Certified Applicators or persons under their direct supervision and only for those uses covered by the Certified Applicator's certification.

NUSYN-NOXFISH®

Fish Toxicant
Synergized Rotenone
Liquid-Emulsifiable

* ***FOR CONTROL OF FISH IN LAKES, PONDS, RESERVOIRS AND STREAMS***

ACTIVE INGREDIENTS:

Rotenone	2.5%
Other Associated Resins	2.5%
Piperonyl Butoxide, Technical*	2.5%
OTHER INGREDIENTS:+	92.5%
	100.0%

* Equivalent to 2.0% [Butylcarbityl] [6-propylpiperonyl] ether and 0.5% related compounds.
+ This product contains aromatic petroleum solvent.
®Nusyn-Noxfish is a registered trademark of AgrEvo Environmental Health, Inc.

EPA Reg. No. 432-550 **EPA EST. NO.**

KEEP OUT OF REACH OF CHILDREN

DANGER

POISONOUS

FIRST AID

IF INHALED: Remove victim to fresh air. If not breathing, give artificial respiration, preferably mouth to mouth. Get medical attention.

IF IN EYES: Hold eyelids open and flush with a steady, gentle stream of water for 15 minutes. Get medical attention.

IF SWALLOWED: Promptly drink a large quantity of milk, egg white, gelatin solution or if these are not available, large quantities of water. Avoid alcohol. Do not induce vomiting. Call a physician or Poison Control Center.

IF ON SKIN: Wash with plenty of soap and water. Get medical attention.

See Side Panel For Additional Precautionary Statements

NET CONTENTS

550s Q080699tk

In case of Medical emergencies or health and safety inquiries or in case of fire, leaking or damaged containers, information may be obtained by calling 1-800-471-0660.

For product information Call Toll-Free:
1-800-331-2867

PRECAUTIONARY STATEMENTS

Hazards to Humans and Domestic Animals

DANGER

Fatal if inhaled. May be fatal if swallowed. Harmful if absorbed through skin. Causes substantial but temporary eye injury. Causes skin irritation. Do not breath spray mist. Do not get in eyes, on skin or on clothing. Wear goggles or safety glasses.

When working with undiluted product wear either a respirator with an organic-vapor removing cartridge with a prefilter approved for pesticides (MSHA/NIOSH approval number prefix TC-23C), or a canister approved for pesticides (MSHA/NIOSH approval number prefix 14G) or a NIOSH approved respirator with an organic vapor (OV) cartridge or canister with any R, P or HE prefilter.

Wash thoroughly with soap and water after handling and before eating, drinking or using tobacco. Remove contaminated clothing and wash before reuse.

Environmental Hazards

This pesticide is extremely toxic to fish. Fish kills are expected at recommended rates. Consult your State Fish and Game Agency before applying this product to public waters to determine if a permit is needed for such an application. Do not contaminate untreated water when disposing of equipment washwaters.

Chemical and Physical Hazards

Combustible mixture. Flash point of this formulation is 115° F. DO NOT USE OR STORE NEAR HEAT OR OPEN FLAME.

DIRECTIONS FOR USE

It is a violation of federal law to use this product in a manner inconsistent with its labeling.

STORAGE AND DISPOSAL

Do not contaminate water, food or feed by storage or disposal.
Storage: Store only in original containers, in a dry place inaccessible to children and pets. Nusyn-Noxfish will not solidify nor show any separation at temperatures down to 40°F and is stable for a minimum of one year when stored in sealed drums at 70°F.
Pesticide Disposal: Pesticide wastes are acutely hazardous. Improper disposal of excess pesticide, spray mixture, or rinsate is a violation of federal law. If these wastes cannot be disposed of by use according to label instructions contact your state pesticide or Environmental Control Agency, or the Hazardous Waste representative at the nearest EPA Regional Office for guidance.
Container Disposal: Triple rinse (or equivalent). Then offer for recycling or reconditioning, or puncture and dispose of in a sanitary landfill, or by other procedures approved by state and local authorities.

General Information

Nusyn-Noxfish is a specially formulated product containing synergized rotenone to be used in fisheries management for the eradication of fish from lakes, ponds, reservoirs and streams.

Nusyn Noxfish is registered for use by or under permit from, and after consultation with State and Federal Fish and wildlife Agencies.

Since such factors as pH, temperature, depth and turbidity will change effectiveness, use this product only at locations, rates, and times authorized and approved by appropriate state and federal fish and wildlife agencies. Rates must be within the range specified on the label.

Properly dispose of dead fish and unused product. Do not use dead fish for food or feed.

Do not use water treated with rotenone to irrigate crops or release within 1\2 mile upstream of a potable water or irrigation water intake in a standing body of water such as a lake, pond or reservoir.

Re-entry Statement: Do not allow swimming in rotenone-treated water until the application has been completed and all pesticide has been thoroughly mixed into the water according to labeling instructions.

For Use in Ponds, Lakes and Reservoirs

The actual application rates and concentrations of rotenone needed to control fish will vary widely, depending on the type of use (e.g. selective treatment, normal pond use, etc.) and the factors listed above. The table below is a general guide for the proper rates and concentrations.

Nusyn-Noxfish disperses readily in water both laterally and vertically, and will penetrate below the thermocline in thermally stratified bodies of water.

Computation of Acre-Feet: An acre-foot is a unit of volume of a body of water having the area of one acre and the depth of one foot. To determine acre feet in a given body of water, make a series of transects across the body of water taking depths with a measured pole or weighted line. Add the soundings and divide by the number made to determine the average depth. Multiply this average depth by the total surface area in order to determine the acre feet to be treated. If the number of surface acres is unknown, contact your local Soil Conservation Service, which can determine this from aerial photographs.

Amount of Nusyn-Noxfish Needed for Specific Uses: To determine the approximate number of gallons of Nusyn-Noxfish (2.5% Rotenone) needed, find your "Type of Use" in the first column of the table below and then divide the corresponding numbers in the fourth column, "Number of Acre-Feet Covered by One Gallon", into the number of acre-feet in your body of water.

Pre-Mixing and Method of Application: Pre-mix with water at a rate of one gallon Nusyn-Noxfish to 10 gallons of water. Uniformly apply over water surface or bubble through underwater lines.

Detoxification: Nusyn-Noxfish treated waters detoxify under natural conditions within one week to one month depending upon temperatures, alkalinity, etc. Rapid detoxification can be accomplished by adding chlorine or potassium permanganate to the water at the same rate as Nusyn-Noxfish in parts per million, plus enough additional to meet the chlorine demand of the untreated water.

Removal of Taste and Odor: Nusyn-Noxfish treated waters do not retain a detectable taste or odor for more than a few days to a maximum of one month. Taste and odor can be removed immediately by treatment with activated charcoal at a rate of 30 ppm for each 1 ppm Nusyn-Noxfish remaining. (Note: As Nusyn-Noxfish detoxifies, less charcoal is required.)

Restocking After Treatment: Wait 2 to 4 weeks after treatment. Place a sample of fish to be stocked in wire cages in the coolest part of the treated waters. If the fish are not killed within 24 hours, the water may be restocked.

Use in Streams Immediately Above Lakes, Ponds and Reservoirs

The purpose of treating streams immediately above lakes, ponds and reservoirs is to improve the effectiveness of lake, pond and reservoir treatments by preventing target fish from moving into the stream corridors and not to control fish in streams per se. The term "immediately" means the first available site above the lake, pond or reservoir where treatment is practical, while still creating a sufficient barrier to prevent migration of target fish into the stream corridor.

In order to completely clear a fresh water aquatic habitat of target fish, the entire system above or between fish barriers must be treated. See the use directions for streams and rivers on this label for proper application instructions.

General Guide to the Application Rates and Concentrations of Rotenone Needed to Control Fih in Lakes, Ponds and Reservoirs[1]

Type of Use	Parts Per Million		Number of Acre-Feet Covered by One Gallon
	Nusyn-Noxfish	Active Rotenone	
Selective Treatment	0.20 to 0.25	0.005 to 0.007	15 to 12
Normal pond use	1.0 to 2.0	0.025 to 0.050	3.0 to 1.5
Remove bullheads or carp	2.0 to 4.0	0.050 to 0.100	1.5 to 0.75
Remove bullheads or carp in rich organic ponds	4.0 to 8.0	0.100 to 0.200	0.75 to 0.38
Preimpoundment treatment above dam	6.0 to 10.0	0.150 to 0.250	0.50 to 0.30

[1]Adapted from Kinney, Edward, 1965. Rotenone in Fish Pond Management. USDI Washington, D.C. Leaflet FL-576.

NUSYN-NOXFISH®

In order to treat a stream immediately above a lake, pond or reservoir, you must: (a) select the concentration of active rotenone, (b) compute the flow rate of the stream, (c) calculate the application rate, (d) select an exposure time, (e) estimate the amount of product needed, (f) follow the method of application.

To prevent movement of fish from the pond, lake or reservoir, stream treatment should begin before and continue throughout treatment of pond, lake or reservoir until mixing has occurred.

1. Concentration of Active Rotenone:
Select the concentration of active rotenone based on the type of use from those listed on the table. Example: If you select "normal pond use", you could select a concentration of 0.025 part per million.

2. Computation of Flow Rate for Stream:
Select a cross section of the stream where the banks and bottom are relatively smooth and free of obstacles. Divide the surface width into 3 equal sections and determine the water depth and surface velocity at the center of each section. In slowly moving streams, determine the velocity by dropping a float attached to 5 feet of loose, monofilament fishing line. Measure the time required for the float to move 5 feet. For fast-moving streams, use a longer distance. Take at least three readings at each point. To calculate the flow rate from the information obtained above, use the following formula:

$$F = \frac{Ws \times D \times L \times C}{T}$$

Where F = flow rate (cubic feet/second), Ws = surface width (feet), D = mean depth (feet), L = mean distance traveled by float (feet), C = constant (0.8 for rough bottoms and 0.9 for smooth bottoms), and T = mean time for float (sec.).

3. Calculation of Application Rate:
In order to calculate the application rate (expressed as gallons/ second), convert the rate in the table (expressed as gallons/ acre-feet), to gallons per cubic feet and multiply by the flow rate (expressed as cubic feet/second). Depending on the size of the stream and the type of equipment, the rate could be expressed in other units, such as ounces/hour, or cc/minute.

The application rate for the stream is calculated as follows:

$$Rs = Rp \times C \times F$$

Where Rs = application rate for stream (gallons/second), Rp = application rate for pond (gallons/acre-feet), C = 1 acre foot/43,560 cubic feet, and F = flow rate of the stream (cubic feet/ second).

4. Exposure Time:
The exposure time would be the period of time (expressed in hours or minutes) during which Nusyn-Noxfish is applied to the steam in order to prevent target fish from escaping from the pond into the stream corridor.

5. Amount of Product:
Calculate the amount of product for a stream by multiplying the application rate for streams by the exposure time.

$$A = Rs \times H$$

Where A = the amount of product for the stream application, Rs = application rate for stream (gallons/second), and H = the exposure time expressed in seconds.

For Use in Streams and Rivers

Only state or federal fish and wildlife personnel or professional fisheries biologists under the authorization of state or federal fish and wildlife agencies are permitted to make applications of Nusyn-Noxfish for control of fish in streams and rivers. Informal consultation with fish and wildlife personnel regarding the potential occurrence of endangered species in areas to be treated should take place. Applicators must reference AgrEvo Environmental Health's Nusyn-Noxfish Stream and River Use Monograph before making any application to streams or rivers.

IMPORTANT: READ BEFORE USE

By using this product, user or buyer accepts the following conditions, disclaimer of warranties and limitations of liability.

CONDITIONS: The directions for use of this product are believed to be adequate and should be followed carefully. However, because of manner of use and other factors beyond AgrEvo Environmental Health's control, it is impossible for AgrEvo Environmental Health to eliminate all risks associated with the use of this product. All such risks shall be assumed by the user or buyer.

DISCLAIMER OF WARRANTIES: THERE ARE NO WARRANTIES, EXPRESS OR IMPLIED, OF MERCHANTABILITY OR OF FITNESS FOR A PARTICULAR PURPOSE OR OTHERWISE, WHICH EXTEND BEYOND THE STATEMENTS MADE ON THIS LABEL. No agent of AgrEvo Environmental Health is authorized to make any warranties beyond those contained herein or to modify the warranties contained herein. AgrEvo Environmental Health disclaims any liability whatsoever for incidental or consequential damages, including, but not limited to, liability arising out of breach of contract, express or implied warranty (including warranties of merchantability and fitness for a particular purpose), tort, negligence, strict liability or otherwise.

LIMITATIONS OF LIABILITY: THE EXCLUSIVE REMEDY OF THE USER OR BUYER FOR ANY AND ALL LOSSES, INJURIES OR DAMAGES RESULTING FROM THE USE OR HANDLING OF THIS PRODUCT, WHETHER IN CONTRACT, WARRANTY, TORT, NEGLIGENCE, STRICT LIABILITY OR OTHERWISE, SHALL NOT EXCEED THE PURCHASE PRICE PAID, OR AT AGREVO ENVIRONMENTAL HEALTH'S ELECTION, THE REPLACEMENT OF PRODUCT.

AgrEvo Environmental Health
95 Chestnut Ridge Road
Montvale, NJ 07645

SPECIMEN LABEL

RESTRICTED USE PESTICIDE
Due to Aquatic, Acute Oral and Inhalation Toxicity

For retail sale to, and use only by, Certified Applicators, or persons under their direct supervision, and only for those uses covered by the Certified Applicator's certification.

PRO-NOXFISH® DUST
Fish Toxicant

ACTIVE INGREDIENTS:

Rotenone*	7.4%
Other Associated Resins	11.1%
INERT INGREDIENTS:	81.5%
Total	100.0%

*Nominal concentration. See certificate of analysis for actual rotenone content.
®PRO-NOXFISH is a registered trademark of AgrEvo Environmental Health, Inc.

EPA REG. NO. 432-829 **EPA EST. NO.**

KEEP OUT OF REACH OF CHILDREN

 **DANGER POISON** 

FIRST AID

IF INHALED: Remove victim to fresh air. If not breathing give artificial respiration preferably mouth-to-mouth. Get medical attention.
IF SWALLOWED: Call a physician or Poison Control Center. Drink 1 to 2 glasses of water and induce vomiting by touching back of throat with finger. If person is unconscious do not give anything by mouth and do not induce vomiting.
IF IN EYES: Flush eyes with plenty of water. Call a physician if irritation persists.
IF ON SKIN: Wash with plenty of soap and water. Get medical attention.

PRECAUTIONARY STATEMENTS
Hazards to Humans and Domestic Animals

DANGER
Fatal if inhaled or swallowed, Harmful if absorbed through skin. Causes moderate eye irritation. Prolonged or frequently repeated skin contact may cause allergic reactions in some individuals. Do not breath dust. Use a dust filtering respirator (MSHA/NIOSH approval number prefix TC-21c). Avoid contact with skin, eyes or clothing. Wash thoroughly with soap and water after handling and before eating, drinking or using tobacco. Remove contaminated clothing and wash clothing before reuse.

Net Contents

Q829pro 102997ln

Environmental Hazards

This pesticide is extremely toxic to fish. Fish kills are expected at recommended rates. Consult your State Fish and Game Agency before applying this product to public waters to determine if a permit is needed for such an application. Do not contaminate untreated water when disposing of equipment washwaters.

DIRECTIONS FOR USE

It is a violation of Federal law to use this product in a manner inconsistent with its labeling.

STORAGE AND DISPOSAL

Do not contaminate water, food or feed by storage or disposal.
Storage: Store only in original containers, in a dry place inaccessible to children and pets.
Pesticide Disposal: Pesticide Wastes are acutely hazardous. Improper disposal of excess pesticide, spray mixture, or rinsate is a violation of Federal Law. If these wastes cannot be disposed of by use according to label instructions, contact your State Pesticide or Environmental Control Agency or the Hazardous Waste Representative at the nearest EPA Regional Office for guidance.
Container Disposal: Completely empty liner by shaking and tapping sides and bottom to loosen clinging particles. Empty residue into application equipment. Then dispose of liner in a sanitary landfill or by incineration if allowed by State and local authorities. If drum is contaminated and cannot be reused, dispose of in the same manner.

USE RESTRICTIONS
Use against fish in lakes, ponds, reservoirs and streams (immediately above lakes, ponds and reservoirs).

Pro-Noxfish Dust Fish Toxicant is registered for use by or under permit from, and after consultation with, State and Federal Fish and Wildlife Agencies.

Since such factors as pH, temperature, depth and turbidity will change effectiveness, use this product only at locations, rates, and times authorized and approved by appropriate state and federal fish and wildlife agencies. Rates must be within the range specified on the label.

Properly dispose of dead fish and unused product. Do not use dead fish for food or feed. Do not use water treated with rotenone to irrigate crops or release within 1/2 mile upstream of a potable water or irrigation water intake in a standing body of water such as a lake, pond or reservoir.

RE-ENTRY STATEMENT
Do not allow swimming in rotenone treated water until the application has been completed and all pesticide has been thoroughly mixed into the water according to labeling instructions.

APPLICATION DIRECTIONS
Treatment of Ponds, Lakes and Reservoirs:
The actual application rates and concentrations of rotenone need-

ed to control fish will vary widely, depending on the type of use (e.g., selective treatment, normal pond use, etc.) and the factors listed above. The table below is a general guide for the proper rates and concentrations.

COMPUTATION OF ACRE-FEET: An acre-foot is a unit of volume of a body of water having the area of one acre and the depth of one foot. To determine acre feet in a given body of water, make a series of transects across the body of water taking depths with a measured pole or weighted line. Add the soundings and divide by the number made to determine the average depth. Multiply this average depth by the total surface area in order to determine the acre feet to be treated. If number of surface acres is unknown, contact your local Soil Conservation Service, which can determine this from aerial photographs.

AMOUNT OF PRODUCT NEEDED FOR TREATMENT: To determine the approximate number of pounds needed for treatment, find your "Type of Use" in the first column of the table below and then divide the corresponding numbers in the third column, "Number of Acre-Feet Covered by One Pound" into the number of acre-feet in your body of water. This will give you the number of pounds of Pro-Noxfish Dust containing 5% rotenone needed for treatment. To correct for the actual rotenone content of the Pro-Noxfish Dust use the following formula;

$$P = \frac{N \times \text{Actual Rotenone Content}}{0.05}$$

Where N= the number of pounds of Pro-Noxfish Dust containing 5% rotenone needed for treatment, P= number of pounds of Pro-Noxfish Dust (actual concentration) needed for treatment.

General Guide to the Application Rates and Concentrations of Rotenone Needed to Control Fish in Lakes, Ponds and Reservoirs[1]

Type of Use	Parts Per Million		Number of Acre-Feet Covered by One Pound (based on 5% rotenone)	Pounds of Pro Noxfish Dust to Cover one Acre-foot (based on 5% rotenone)
	5% Rotenone	Active Rotenone		
Selective Treatment	0.10 to 0.13	0.005 to 0.007	3.7 to 2.8	0.25 to 0.36
Normal pond use	0.5 to 1.0	0.025 to 0.050	0.74 to 0.37	1.35 to 2.70
Remove bullheads or carp	1.0 to 2.0	0.050 to 0.100	0.37 to 0.185	2.70 to 5.41
Remove bullheads or carp in rich organic ponds	2.0 to 4.0	0.100 to 0.200	0.185 to 0.093	5.41 to 10.75
Preimpoundment treatment above dam	3.0 to 5.0	0.150 to 0.250	0.123 to 0.074	8.13 to 13.51

[1]Adapted from Kinney, Edward. 1965. Rotenone in Fish Pond Management. USDI Washington, D.C. Leaflet FL-576.

Pre-Mix and Method of Application: Pre-mix one pound Rotenone with 3 to 10 gallons of water. Uniformly apply over water surface or bubble through underwater lines.

Detoxification: Rotenone treated waters detoxify under natural conditions within one week to one month depending upon temperatures, alkalinity, etc. Rapid detoxification can be accomplished by adding chlorine or potassium permanganate to the water at the same rate as Rotenone in parts per million, plus enough additional to meet the chlorine demand of the untreated water.

Removal of Taste and Odor: Rotenone treated waters do not retain a detectable taste or odor for more than a few days to a maximum of one month. Taste and odor can be removed immediately by treatment with activated charcoal at a rate of 30 ppm for each 1 ppm Rotenone. (Note: As Rotenone detoxifies, less charcoal is required.)

Restocking After Treatment: Wait 2 to 4 weeks after treatment. Place a sample of fish to be stocked in wire cages in the coolest part of the treated waters. If the fish are not killed within 24 hours, the water may be restocked.

Use in Streams Immediately Above Lakes, Ponds and Reservoirs: The purpose of treating streams immediately above lakes, ponds and reservoirs is to improve the effectiveness of lake, pond and reservoir treatments by preventing target fish from moving into the stream corridors, and not to control fish in streams per se. The term "immediately" means the first available site above the lake, pond or reservoir where treatment is practical, while still creating a sufficient barrier to prevent migration of target fish into the stream corridor.

In order to completely clear a fresh water aquatic habitat of target fish, the entire system above or between fish barriers must be treated. See the use directions for streams and rivers on this label for proper application instructions.

In order to treat a stream immediately above a lake, pond or reservoir, you must: (a) select the concentration of active rotenone, (b) compute the flow rate of the stream, (c) calculate the application rate, (d) select an exposure time, (e) estimate the amount of product needed, (f) follow the method of application. To prevent movement of fish from the pond, lake or reservoir, stream treatment should begin before and continue throughout treatment of pond, lake or reservoir until mixing has occurred.

1. Concentration of Active Rotenone: Select the concentration of active rotenone based on the type of use from those listed on the table. Example: If you select "normal pond use" you could select a concentration of 0.025 part per million.

2. Computation of Flow Rate for Stream: Select a cross section of the stream where the banks and bottom are relatively smooth and free of obstacles. Divide the surface width into 3 equal sections and determine the water depth and surface velocity at the center of each section. In slowly moving streams, determine the velocity by dropping a float attached to 5 feet of loose monofilament fishing line. Measure the time required for the float to move 5 feet. For fast-moving streams, use a longer distance. Take at least three readings at each point. To calculate the flow rate from the information obtained above, use the following formula:

$$F = \frac{Ws \times D \times L \times C}{T}$$

PRO-NOXFISH® DUST

where F = flow rate (cubic feet/second), Ws = surface width (feet), D = mean depth (feet), L = mean distance traveled by float (feet), C = constant (0.8 for rough bottoms and 0.9 for smooth bottoms), and T = mean time for float (sec.).

3. Calculation of Application Rate: In order to calculate the application rate (expressed as pounds/second), you convert the rate in the table (expressed as pounds/acre-feet), to gallons per cubic feet and multiply by the flow rate (expressed as cubic feet/second). Depending on the size of the stream and the type of equipment, the rate could be expressed in other units, such as ounces/hour.

The application rate for the stream is calculated as follows:

$$R_s = Rp \times C \times F$$

where Rs = application rate for stream (pounds/second), Rp = application rate for pond (pounds/acre-feet), C = 1 acre foot/43560 cubic feet, and F = flow rate of the stream (cubic feet/second).

4. Exposure Time: The exposure time would be the period of time (expressed in hours or minutes) during which Rotenone is applied to the stream in order to prevent target fish from escaping from the pond into the stream corridor.

5. Amount of Product: Calculate the amount of product for a stream by multiplying the application rate for streams by the exposure time.

$$A = Rs \times H$$

where A = the amount of product for the stream application, Rs = application rate for stream (pounds/second), and H = the exposure time expressed in seconds.

WARRANTY STATEMENT

Our recommendations for the use of this product are based upon tests believed to be reliable. The use of this product being beyond the control of the manufacturer, no guarantee, expressed or implied, is made as to the effects of such or the results to be obtained if not used in accordance with directions or established safe practice. The buyer must assume all responsibility including injury or damage, resulting from its misuse as such, or in combination with other materials.

AgrEvo Environmental Health
95 Chestnut Ridge Road
Montvale, NJ 07645

Product of Peru

SPECIMEN LABEL

RESTRICTED USE PESTICIDE

DUE TO AQUATIC AND ACUTE INHALATION TOXICITY

For retail sale to, and use only by, Certified Applicators or persons under their direct supervision and only for those uses covered by the Certified Applicator's certification.

PRENFISH TOXICANT

Liquid Emulsifiable

*For Control of Fish in Lakes, Ponds, Reservoirs and Streams

ACTIVE INGREDIENTS:	
Rotenone	5.0%
Other Associated Resins	10.0%
INERT INGREDIENTS*:	85.0%
TOTAL	100.0%

*This product contains aromatic hydrocarbons.
PRENTOX® - Registered Trademark of Prentiss Incorporated

KEEP OUT OF REACH OF CHILDREN

DANGER - POISONOUS

See Left Panel for additional precautionary statements.

EPA Reg. No. 655-422

EPA Est. No. 655-GA-1

Manufactured by:

PRENTISS INCORPORATED

Plant: Kaolin Road, Sandersville, GA 31082
Office: C.B. 2000, Floral Park, NY 11002-2000

Page 1 of 4

SPECIMEN LABEL

PRECAUTIONARY STATEMENTS
HAZARDS TO HUMANS AND DOMESTIC ANIMALS
DANGER

Fatal if inhaled. May be fatal if swallowed. Harmful if absorbed through skin. Causes substantial but temporary eye injury. Causes skin irritation. Do not breath spray mist. Do not get in eyes, on skin or on clothing. Wear goggles or safety glasses.

Wear either a respirator with an organic-vapor-removing cartridge with a prefilter approved for pesticides (MSHA/NIOSH approval number prefix TC-23C), or a canister approved for pesticides (MSHA/NIOSH approval number prefix 14G), or a NIOSH approved respirator with an organic vapor (OV) cartridge or canister with any R, P or HE prefilter.

Wash thoroughly with soap and water after handling and before eating, drinking or using tobacco. Remove contaminated clothing and wash before reuse.

STATEMENT OF PRACTICAL TREATMENT

If inhaled: Remove victim to fresh air. If not breathing, give artificial respiration, preferably mouth-to-mouth. Get medical attention.
If in eyes: Hold eyelids open and flush with a steady, gentle stream of water for 15 minutes. Get medical attention.
If swallowed: Promptly drink a large quantity of milk, egg white, gelatin solution, or if these are not available, large quantities of water. Avoid alcohol. Do not induce vomiting. Call a physician or Poison Control Center.
If on skin: Wash with plenty of soap and water. Get medical attention.

ENVIRONMENTAL HAZARDS

This pesticide is extremely toxic to fish. Fish kills are expected at recommended rates. Consult your State Fish and Game Agency before applying this product to public waters to determine if a permit is needed for such an application. Do not contaminate untreated water when disposing of equipment washwaters.

CHEMICAL AND PHYSICAL HAZARDS

FLAMMABLE: KEEP AWAY FROM HEAT AND OPEN FLAME. FLASH POINT MINIMUM 45° F (7° C).

STORAGE AND DISPOSAL

Do not contaminate water, food or feed by storage or disposal.
Storage: Store only in original containers, in a dry place inaccessible to children and pets. Prentox Prenfish Toxicant will not solidify nor show any separation at temperatures down to 40°F and is stable for a minimum of one year when stored in sealed drums at 70°F.
Pesticide Disposal: Pesticide wastes are acutely hazardous. Improper disposal of excess pesticide, spray mixture, or rinsate is a violation of federal law. If these wastes cannot be disposed of by use according to label instructions contact your state pesticide or Environmental Control Agency, or the Hazardous Waste representative at the nearest EPA Regional Office for guidance.
Container Disposal: Triple rinse (or equivalent). Then offer for recycling or reconditioning, or puncture and dispose of in a sanitary landfill, or by other procedures approved by state and local authorities.

DIRECTIONS FOR USE

It is a violation of Federal law to use this product in a manner inconsistent with its labeling.

General Information

Prentox Prenfish Toxicant is a specially formulated product containing rotenone, to be used in fisheries management for the eradication of fish from lakes, ponds, reservoirs and streams.

Since such factors as pH, temperature, depth and turbidity will change effectiveness, use this product only at locations, rates, and times authorized and approved by appropriate state and federal fish and wildlife agencies. Rates must be within the range specified on the label.

Properly dispose of unused product. Do not use dead fish for food or feed.

Do not use water treated with rotenone to irrigate crops or release within 1/2 mile upstream of a potable water or irrigation water intake in a standing body of water such as a lake, pond or reservoir.

Re-entry Statement: Do not allow swimming in rotenone-treated water until the application has been completed and all pesticide has been thoroughly mixed into the water according to labeling instructions.

SPECIMEN LABEL

For Use in Ponds, Lakes and Reservoirs
The actual application rates and concentrations of rotenone needed to control fish will vary widely, depending on the type of use (e.g., selective treatment, normal pond use, etc.) and the factors listed above. The table below is a general guide for the proper rates and concentrations.

Prentox Prenfish Toxicant disperses readily in water both laterally and vertically, and will penetrate below the thermocline in thermally stratified bodies of water.

Computation of Acre-Feet: An acre-foot is a unit of volume of a body of water having the area of one acre and the depth of one foot. To determine acre feet in a given body of water, make a series of transects across the body of water taking depths with a measured pole or weighted line. Add the soundings and divide by the number made to determine the average depth. Multiply this average depth by the total surface area in order to determine the acre feet to be treated. If number of surface acres is unknown, contact your local Soil Conservation Service, which can determine this from aerial photographs.

Amount of Prentox Prenfish Toxicant Needed for Specific Uses: To determine the approximate number of gallons of Prentox Prenfish Toxicant (5.0% Rotenone) needed, find your "Type of Use" in the first column of the table below and then divide the corresponding numbers in the forth column, "Number of Acre-Feet Covered by One Gallon" into the number of acre-feet in your body of water.

General Guide to the Application Rates and Concentrations of Rotenone Needed to Control Fish in Lakes, Ponds and Reservoirs[1]

Type of Use	Parts Per Million Prenfish Toxicant	Parts Per Million Active Rotenone	Number of Acre-Feet Covered by One Gallon
Selective Treatment	0.10 to 0.13	0.005 to 0.007	30 to 24
Normal Pond Use	0.5 to 1.0	0.025 to 0.050	6.0 to 3.0
Remove bullheads or carp	1.0 to 2.0	0.050 to 0.100	3.0 to 1.5
Remove bullheads or carp in rich organic ponds	2.0 to 4.0	0.100 to 0.200	1.5 to 0.75
Preimpoundment treatment above dam	3.0 to 5.0	0.150 to 0.250	1.0 to 0.60

[1] Adapted from Kinney, Edward. 1965. Rotenone in Fish Pond Management. USDI Washington, D.C. Leaflet FL-576.

Pre-Mixing and Method of Application: Pre-mix with water at a rate of one gallon Prentox Prenfish Toxicant to 10 gallons of water. Uniformly apply over water surface or bubble through underwater lines.

Detoxification: Prentox Prenfish Toxicant treated waters detoxify under natural conditions within one week to one month depending upon temperatures, alkalinity, etc. Rapid detoxification can be accomplished by adding chlorine or potassium permanganate to the water at the same rate as Prentox Prenfish Toxicant in parts per million, plus enough additional to meet the chlorine demand of the untreated water.

Removal of Taste and Odor: Prentox Prenfish Toxicant treated waters do not retain a detectable taste or odor for more than a few days to a maximum of one month. Taste and odor can be removed immediately by treatment with activated charcoal at a rate of 30 ppm for each 1 ppm Prentox Prenfish Toxicant remaining. (Note: As Prentox Prenfish Toxicant detoxifies, less charcoal is required.)

Restocking After Treatment: Wait 2 to 4 weeks after treatment. Place a sample of fish to be stocked in wire cages in the coolest part of the treated waters. If the fish are not killed within 24 hours, the water may be restocked.

Use in Streams Immediately Above Lakes, Ponds and Reservoirs

The purpose of treating streams immediately above lakes, ponds and reservoirs is to improve the effectiveness of lake, pond and reservoir treatments by preventing target fish from moving into the stream corridors, and not to control fish in streams per se. The term "immediately" means the first available site above the lake, pond or reservoir where treatment is practical, while still creating a sufficient barrier to prevent migration of target fish into the stream corridor.

In order to completely clear a fresh water aquatic habitat of target fish, the entire system above or between fish barriers must be treated. See the use directions for streams and rivers on this label for proper application instructions.

In order to treat a stream immediately above a lake, pond or reservoir you must: (a) select the concentration of active rotenone, (b) compute the flow rate of the stream, (c) calculate the application rate, (d) select an exposure time, (e) estimate the amount of product needed, (f) follow the method of application. To prevent movement of fish from the pond, lake or reservoir, stream treatment should begin before and continue throughout treatment of the pond, lake or reservoir until mixing has occurred.

SPECIMEN LABEL

1. Concentration of Active Rotenone
Select the concentration of active rotenone based on the type of use from those listed on the table. Example: If you select "normal pond use" you could select a concentration of 0.025 part per million.

2. Computation of Flow Rate for Stream
Select a cross section of the stream where the banks and bottom are relatively smooth and free of obstacles. Divide the surface width into 3 equal sections and determine the water depth and surface velocity at the center of each section. In slowly moving streams, determine the velocity by dropping a float attached to 5 feet of loose monofilament fishing line. Measure the time required for the float to move 5 feet. For fast-moving streams, use a longer distance. Take at least three readings at each point. To calculate the flow rate from the information obtained above, use the following formula:

$$F = \frac{Ws \times D \times L \times C}{T}$$

Where F = flow rate (cubic feet/second), Ws = surface width (feet), D = mean depth (feet), L = mean distance traveled by float (feet), C = constant (0.8 for rough bottoms and 0.9 for smooth bottoms), and T = mean time for float (sec.).

3. Calculation of Application Rate
In order to calculate the application rate (expressed as gallons/second), you convert the rate in the table (expressed as gallons/acre-feet), to gallons per cubic feet and multiply by the flow rate (expressed as cubic feet/second). Depending on the size of the stream and the type of equipment, the rate could be expressed in other units, such as ounces/hour, or cc/minute.

The application rate for the stream is calculated as follows:

$$R_s = R_p * C * F$$

where R_s = application rate for stream (gallons/second), R_p = application rate for pond (gallons/acre-feet), C = 1 acre foot/43560 cubic feet, and F = flow rate of the stream (cubic feet/second).

4. Exposure Time
The exposure time would be the period of time (expressed in hours or minutes) during which Prentox Prenfish Toxicant is applied to the stream in order to prevent target fish from escaping from the pond into the stream corridor.

5. Amount of Product
Calculate the amount of product for a stream by multiplying the application rate for streams by the exposure time.

$$A = R_s * H$$

where A = the amount of product for the stream application, R_s = application rate for stream (gallons/second), and H = the exposure time expressed in seconds.

For use in Streams and Rivers

Only state or federal Fish and Wildlife personnel or professional fisheries biologists under the authorization of state or federal Fish and Wildlife Agencies are permitted to make applications of Prentox Prenfish Toxicant for control of fish in streams and rivers. Informal consultation with Fish and Wildlife personnel regarding the potential occurrence of endangered species in areas to be treated should take place. Applicators must reference Prentiss Incorporated's Prentox Prenfish Toxicant Stream and River Use Monograph before making any application to streams or rivers.

Warranty Statement: Our recommendations for the use of this product are based upon tests believed to be reliable. The use of this product being beyond the control of the manufacturer, no guarantee, expressed or implied, is made as to the effects of such or the results to be obtained if not used in accordance with directions or established safe practice. The buyer must assume all responsibility, including injury or damage, resulting from its misuse as such, or in combination with other materials.

10/98

SPECIMEN LABEL

RESTRICTED USE PESTICIDE

DUE TO AQUATIC AND ACUTE INHALATION TOXICITY

For retail sale to, and use only by, Certified applicators or persons under their direct supervision and only for those uses covered by the Certified Applicator's certification.

SYNPREN-FISH TOXICANT

Liquid-Emulsifiable

*For Control of Fish in Lakes, Ponds, Reservoirs and Streams

ACTIVE INGREDIENTS:	
Rotenone	2.5% w/w
Other Associated Resins	5.0%
Piperonyl Butoxide, Technical*	2.5%
INERT INGREDIENTS:**	90.0%
TOTAL:	100.0%

*Equivalent to 2.0% [Butylcarbityl] [6-propylpiperonyl] ether and 0.5% related compounds.
**This product contains aromatic petroleum solvents.
PRENTOX® - Registered Trademark of Prentiss Incorporated

KEEP OUT OF REACH OF CHILDREN

DANGER - POISONOUS

See Additional Precautionary Statements Below.

E.P.A. REG. NO. 655-421 10/95a E.P.A. EST. NO. 655-GA-1

Manufactured by:

PRENTISS INCORPORATED

Plant: Kaolin Road, Sandersville, GA 31082
Office: C.B. 2000, Floral Park, NY 11002-2000

SPECIMEN LABEL

PRECAUTIONARY STATEMENTS
HAZARDS TO HUMANS AND DOMESTIC ANIMALS
DANGER

Fatal if inhaled. May be fatal if swallowed. Harmful if absorbed through skin. Causes substantial but temporary eye injury. Causes skin irritation. Do not breathe spray mist. Do not get in eyes, on skin or on clothing. Wear goggles or safety glasses.

Wear either a respirator with an organic-vapor-removing cartridge with a prefilter approved for pesticides (MSHA/NIOSH approval number prefix TC-23C), or a canister approved for pesticides (MSHA/NIOSH approval number prefix TC-14G), or a NIOSH approved respirator with an organic vapor (OV) cartridge or canister with any R, P or HE prefilter.

Wash thoroughly with soap and water after handling and before eating, drinking or using tobacco. Remove contaminated clothing and wash before reuse.

STATEMENT OF PRACTICAL TREATMENT

If inhaled: Remove victim to fresh air. If not breathing, give artificial respiration, preferably mouth to mouth. Get medical attention.
If in eyes: Hold eye lids open and flush with a steady, gentle stream of water for 15 minutes. Get medical attention.
If swallowed: Call a physician or Poison Control Center. Drink promptly a large quantity of milk, egg white, gelatin solution, or if these are not available, large quantities of water. Avoid alcohol. Do not induce vomiting.
If on skin: Wash with plenty of soap and water. Get medical attention.

ENVIRONMENTAL HAZARDS

This pesticide is extremely toxic to fish. Fish kills are expected at recommended rates. Consult your State Fish and Game Agency before applying this product to public waters to determine if a permit is needed for such an application. Do not contaminate untreated water when disposing of equipment washwaters.

CHEMICAL AND PHYSICAL HAZARDS

Combustible mixture. Flash point of this formulation is 115° F. DO NOT USE OR STORE NEAR HEAT OR OPEN FLAME.

STORAGE AND DISPOSAL

Do not contaminate water, food or feed by storage or disposal.

Storage: Store only in original containers, in a dry place inaccessible to children and pets. Prentox Synpren-Fish Toxicant will not solidify nor show any separation at temperatures down to 40° F and is stable for a minimum of one year when stored in sealed drums at 70° F.
Pesticide Disposal: Pesticide wastes are acutely hazardous. Improper disposal of excess pesticide, spray mixture, or rinsate is a violation of Federal law. If these wastes cannot be disposed of by use according to label instructions contact your state pesticide or Environmental Control Agency, or the Hazardous Waste representative at the nearest EPA Regional Office for guidance.
Container Disposal: Triple rinse (or equivalent). Then offer for recycling or reconditioning, or puncture and dispose of in a sanitary landfill, or by other procedures approved by state and local authorities.

DIRECTIONS FOR USE

It is a violation of Federal law to use this product in a manner inconsistent with its labeling.

General Information
Prentox Synpren-Fish Toxicant is a specially formulated product containing synergized rotenone, to be used in fisheries management for the eradication of fish from lakes, ponds, reservoirs and streams.

Since such factors as pH, temperature, depth and turbidity will change effectiveness, use this product only at locations, rates, and times authorized and approved by appropriate state and federal fish and wildlife agencies. Rates must be within the range specified on the label.

Properly dispose of unused product. Do not use dead fish for food or feed.

Do not use water treated with rotenone to irrigate crops or release within 1/2 mile upstream of a potable water or irrigation water intake in a standing body of water such as a lake, pond or reservoir.

RE-ENTRY STATEMENT: Do not allow swimming in rotenone-treated water until the application has been completed and all pesticide has been thoroughly mixed into the water according to labeling instructions.

For Use in Ponds, Lakes and Reservoirs
The actual application rates and concentrations of rotenone needed to control fish will vary widely, depending on the type of use (e.g., selective treatment, normal pond use, etc.) and the factors listed above. The table below is a general guide for the proper rates and concentrations.

Prentox Synpren-Fish Toxicant disperses readily in water both laterally and vertically, and will penetrate below the thermocline in thermally stratified bodies of water.

Page 2 of 4

SPECIMEN LABEL

Computation of Acre-Feet: An acre-foot is a unit of volume of a body of water having the area of one acre and the depth of one foot. To determine acre feet in a given body of water, make a series of transects across the body of water taking depths with a measured pole or weighted line. Add the soundings and divide by the number made to determine the average depth. Multiply this average depth by the total surface area in order to determine the acre-feet to be treated. If number of surface acres is unknown, contact your local Soil Conservation Service, which can determine this from aerial photographs.

Amount of Prentox Synpren-Fish Toxicant Needed for Specific Uses: To determine the approximate number of gallons of Prentox Synpren-Fish Toxicant (2.5% Rotenone) needed, find your "Type of Use" in the first column of the table below, and then divide the corresponding numbers in the third column, "Number of Acre-Feet Covered by One Gallon" into the number of acre-feet in your body of water.

General Guide to the Application Rates and Concentrations of Rotenone Needed to Control Fish in Lakes, Ponds and Reservoirs[1]

Type of Use	Parts Per Million		Number of Acre-Feet Covered by One Gallon
	Synpren-Fish Toxicant	Active Rotenone	
Selective Treatment	0.20 to 0.25	0.005 to 0.007	15 to 12
Normal Pond Use	1.0 to 2.0	0.025 to 0.050	3.0 to 1.5
Remove bullheads or carp	2.0 to 4.0	0.050 to 0.100	1.5 to 0.75
Remove bullheads or carp in rich organic ponds	4.0 to 8.0	0.100 to 0.200	0.75 to 0.38
Preimpoundment treatment above dam	6.0 to 10.0	0.150 to 0.250	0.50 to 0.30

[1]Adapted from Kinney, Edward. 1965. Rotenone in Fish Pond Management. USDI Washington, D.C. Leaflet FL-576.

Pre-Mix and Method of Application: Pre-mix with water at a rate of one gallon Prentox Synpren-Fish Toxicant to 10 gallons of water. Uniformly apply over water surface or bubble through underwater lines.

Detoxification: Prentox Synpren-Fish Toxicant treated waters detoxify under natural conditions within one week to one month depending upon temperatures, alkalinity, etc. Rapid detoxification can be accomplished by adding chlorine or potassium permanganate to the water at the same rate as Prentox Synpren-Fish Toxicant in parts per million, plus enough additional to meet the chlorine demand of the untreated water.

Removal of Taste and Odor: Prentox Synpren-Fish Toxicant treated waters do not retain a detectable taste or odor for more than a few days to a maximum of one month. Taste and odor can be removed immediately by treatment with activated charcoal at a rate of 30 ppm for each 1 ppm Prentox Synpren-Fish Toxicant remaining. (Note: As Prentox Synpren-Fish Toxicant detoxifies, less charcoal is required.)

Restocking After Treatment: Wait 2 to 4 weeks after treatment. Place a sample of fish to be stocked in wire cages in the coolest part of the treated waters. If the fish are not killed within 24 hours, the water may be restocked.

Use in Streams Immediately Above Lakes, Ponds, and Reservoirs

The purpose of treating streams immediately above lakes, ponds and reservoirs is to improve the effectiveness of lake, pond and reservoir treatments by preventing target fish from moving into the stream corridors, and not to control fish in streams per se. The term "immediately" means the first available site above the lake, pond or reservoir where treatment is practical, while still creating a sufficient barrier to prevent migration of target fish into the stream corridor.

In order to completely clear a fresh water aquatic habitat of target fish, the entire system above or between fish barriers must be treated. See the use directions for streams and rivers on this label for proper application instructions.

In order to treat a stream immediately above a lake, pond or reservoir, you must: (a) select the concentration of active rotenone, (b) compute the flow rate of the stream, (c) calculate the application rate, (d) select an exposure time, (e) estimate the amount of product needed, (f) follow the method of application. To prevent movement of fish from the pond, lake or reservoir, stream treatment should begin before and continue throughout treatment of pond, lake or reservoir until mixing has occurred.

1. Concentration of Active Rotenone:
Select the concentration of active rotenone based on the type of use from those listed on the table. Example: If you select "normal pond use" you could select a concentration of 0.025 part per million.

2. Computation of Flow Rate for Stream:
Select a cross section of the stream where the banks and bottom are relatively smooth and free of obstacles. Divide the surface width into 3 equal sections and determine the water depth and surface velocity at the center of each section. In slowly moving streams, determine the velocity by dropping a float attached to 5 feet of loose, monofilament fishing line. Measure the time required for the float to move 5 feet. For fast-moving streams, use a longer distance. Take at least three readings at each point. To calculate the flow rate from the information obtained above, use the following formula:

SPECIMEN LABEL

$$F = \frac{Ws \times D \times L \times C}{T}$$

where F = flow rate (cubic feet/second), Ws = surface width (feet), D = mean depth (feet), L = mean distance traveled by float (feet), C = constant (0.8 for rough bottoms and 0.9 for smooth bottoms), and T = mean time for float (sec.).

3. Calculation of Application Rate:
In order to calculate the application rate (expressed as gallons/second), you convert the rate in the table (expressed as gallons/acre-feet), to gallons per cubic feet and multiply by the flow rate (expressed as cubic feet/second). Depending on the size of the stream and the type of equipment, the rate could be expressed in other units, such as ounces/hour, or cc/minute.

The application rate for the stream is calculated as follows:

$$R_s = R_p * C * F$$

where R_s = application rate for stream (gallons/second), R_p = application rate for pond (gallons/acre-feet), C = 1 acre foot/43560 cubic feet, and F = flow rate of the stream (cubic feet/second).

4. Exposure Time:
The exposure time would be the period of time (expressed in hours or minutes) during which Prentox Synpren-Fish Toxicant is applied to the stream in order to prevent target fish from escaping from the pond into the stream corridor.

5. Amount of Product:
Calculate the amount of product for a stream by multiplying the application rate for streams by the exposure time.

$$A = R_s * H$$

where A = the amount of product for the stream application, R_s = application rate for stream (gallons/second), and H = the exposure time expressed in seconds.

For Use in Streams and Rivers

Only state or federal Fish & Wildlife personnel or professional fisheries biologists under the authorization of state or federal Fish & Wildlife agencies are permitted to make applications of Prentox Synpren-Fish Toxicant for control of fish in streams and rivers. Informal consultation with Fish & Wildlife personnel regarding the potential occurrence of endangered species in areas to be treated should take place. Applicators must reference Prentiss Incorporated's Prentox Synpren-Fish Toxicant Stream and River Use Monograph before making any application to streams or rivers.

Warranty Statement: Our recommendations for the use of this product are based upon tests believed to be reliable. The use of this product being beyond the control of the manufacturer, no guarantee, expressed or implied, is made as to the effects of such or the results to be obtained if not used in accordance with directions or established safe practice. The buyer must assume all responsibility, including injury or damage, resulting from its misuse as such, or in combination with other materials.

SPECIMEN LABEL

RESTRICTED USE PESTICIDE
DUE TO AQUATIC, ACUTE ORAL AND INHALATION TOXICITY
For retail sale to, and use by, Certified Applicators or persons under their direct supervision and only for those uses covered by the Certified Applicator's certification.

ROTENONE FISH TOXICANT POWDER

ACTIVE INGREDIENTS:

Rotenone- Minimum Guaranteed	7.4% w/w
Other Associated Resins	11.1%
INERT INGREDIENTS:	81.5%
TOTAL:	100.0% w/w

ROTENONE ASSAY ____________________% ROTENONE

PRENTOX® - Registered Trademark of Prentiss Incorporated

KEEP OUT OF REACH OF CHILDREN

DANGER
POISON

SEE INSIDE LEAFLET FOR ADDITIONAL PRECAUTIONARY STATEMENTS

Manufactured by:

PRENTISS INCORPORATED

E.P.A. REG. NO. 655-691
E.P.A. EST. NO. 655-GA-1

Plant: Kaolin Road, Sandersville, GA 31082
Office: C.B. 2000, Floral Park, NY 11002-2000

PRECAUTIONARY STATEMENTS
HAZARDS TO HUMANS AND DOMESTIC ANIMALS
DANGER

Fatal if inhaled or swallowed. Harmful if absorbed through the skin. Causes moderate eye irritation. Prolonged or frequently repeated skin contact may cause allergic reactions in some individuals. Do not breathe dust. Use a dust/mist filtering respirator (MSHA/NIOSH approval number prefix TC-21C), or a NIOSH approved respirator with any N, R, P or HE filter. Avoid contact with skin, eyes or clothing. Wash thoroughly with soap and water after handling and before eating, drinking or using tobacco. Remove contaminated clothing and wash clothing before reuse.

FIRST AID

If inhaled — Remove victim to fresh air. If not breathing, give artificial respiration preferably mouth-to-mouth. Get medical attention. **If swallowed** — Call a physician or Poison Control Center. Drink 1 or 2 glasses of water and induce vomiting by touching back of throat with finger. Do not induce vomiting or give anything by mouth to an unconscious person. **If in eyes** — Flush with plenty of water. Call a physician if irritation persists. **If on skin** — Wash with plenty of soap and water. Get medical attention

ENVIRONMENTAL HAZARDS

This pesticide is extremely toxic to fish. Fish kills are expected at recommended rates. Consult your State Fish and Game Agency before applying this product to public waters to determine if a permit is needed for such an application. Do not contaminate untreated water when disposing of equipment washwaters.

STORAGE AND DISPOSAL

Do not contaminate water, food or feed by storage or disposal.
STORAGE: Store only in original container, in a dry place inaccessible to children and pets. If spilled, sweep up and dispose of as below.
PESTICIDE DISPOSAL: Wastes resulting from the use of this product may be disposed of on site or at an approved waste disposal facility.
CONTAINER DISPOSAL: Completely empty bag into application equipment. Then dispose of bag in a sanitary landfill or by incineration, or if allowed by State and local authorities by burning. If burned, stay out of smoke.

SPECIMEN LABEL

DIRECTIONS FOR USE

It is a violation of Federal law to use this product in a manner inconsistent with its labeling.

USE RESTRICTIONS:

Use against fish in lakes, ponds, and streams (immediately above lakes and ponds).

Since such factors as pH, temperature, depth, and turbidity will change effectiveness, use this product only at locations, rates, and times authorized and approved by appropriate state and Federal fish and wildlife agencies. Rates must be within the range specified in the labeling.

Properly dispose of dead fish and unused product. Do not use dead fish as food or feed.

Do not use water treated with rotenone to irrigate crops or release within ½ mile upstream of a potable water or irrigation water intake in a standing body of water such as a lake, pond or reservoir.

Note to User: Adjust pounds of Rotenone according to the actual Rotenone Assay as noted under the Ingredient Statement on this label. For example, if the required amount of 5% rotenone is 21 pounds, and the Rotenone Assay is 7%, use 5/7 of 21 pounds or 15 pounds of this product to yield the proper amount of active rotenone.

APPLICATION DIRECTIONS:

Treatment of Lakes and Ponds

1. **Application Rates and Concentrations of Rotenone**

The actual application rates and concentrations of rotenone needed to control fish will vary widely, depending on the type of use (e.g. selective treatment, normal pond treatment, etc.) and the factors listed above. The table below is a general guide for the proper rates and concentrations.

2. **Total Amount of Product Needed for Treatment**

To determine the total number of pounds needed for treatment, divide the number of acre-feet covered by one pound for a specific type of use (e.g., selective treatment, etc.), as indicated in the table below, into the number of acre-feet in the body of water.

General Guide to the Application Rates and Concentrations of Rotenone Needed to Control Fish in Lakes and Ponds[1]

Type of Use	No. of Acre-Feet Covered by One Pound	Parts Per Million	
		Active Rotenone	5% Product
Selective Treatment	3.7 to 2.8	0.005 - 0.007	0.10 - 1.3
Normal Pond Use	0.74 to 0.37	0.025 - 0.050	0.5 - 1.0
Remove Bullheads or Carp	0.37 to 0.185	0.050 - 0.100	1.02 - 2.0
Remove Bullheads or Carp in Rich Organic Ponds	0.185 to 0.093	0.100 - 0.200	2.0 - 4.0
Pre-impoundment Treatment above Dam	0.123 to 0.074	0.150 - 0.250	3.0 - 5.0

[1]Adapted from Kinney, Edward, 1965 Rotenone in Fish Pond Management. USDI Washington, D.C. Leaflet FL-576.

Computation of acre-feet for lake or pond: An acre-foot is a unit of water volume having a surface area of one acre and a depth of one foot. Make a series of transects across the surface, taking depths with a measured pole or weighted line. Add the measurements and divide by the number made to determine the average depth. To compute total acre-feet, multiply this average depth by the number of surface acres, which can be determined from an aerial photograph or plat drawn to scale.

3. **Pre-Mixing Method of Application**

Pre-mix one pound of Rotenone with 3 to 10 gallons of water. Uniformly apply over water surface or bubble through underwater lines.

Alternately place undiluted powder in burlap sack and trail behind boat. When treating deep water (20 to 25 feet) weight bag and tow at desired depth.

4. **Removal of Taste and Odor**

Rotenone treated waters do not retain a detectable taste or odor for more than a few days to a maximum of one month. Taste and odor can be removed immediately by treatment with activated charcoal at a rate of 30 ppm. for each 1 ppm. Rotenone remaining (Note: As Rotenone detoxifies, less charcoal is required).

5. **Restocking**

Waters treated with this product detoxify within 2 to 4 weeks after treatment, depending on pH, temperature, water hardness, and depth. To determine if detoxification has occurred, place live boxes containing samples of fish to be stocked in treated waters. More rapid detoxification can be accomplished by adding Potassium Permanganate or chlorine at a 1:1 ratio with the concentration of rotenone applied, plus sufficient additional compound to satisfy the chemical oxidation demand caused by organic matter that may be present in the treated water.

Treatment of Streams Immediately Above Lakes and Ponds

The purpose of treating streams immediately above lakes and ponds is to improve the effectiveness of lake and pond treatments and not to control fish in streams **per se**. The term "immediately" means the first available site above the lake or pond where treatment is practical.

In order to treat a stream immediately above a lake or pond, you must select a concentration of active rotenone, compute the flow rate of a stream, calculate the application rate, select an exposure time, estimate the amount of product needed, and follow the method of application.

SPECIMEN LABEL

1. **Concentration of Active Rotenone**

Select the "Concentration of Active Rotenone" based on the type of use from those on the table. For example, if you select "Normal Pond Use" you could select a concentration of "0.025 Parts per Million".

2. **Computation of Flow Rate for Stream**

Select a cross section of the stream where the banks and bottom are relatively smooth and free of obstacles. Divide the surface width into 3 equal sections and determine the water depth and surface velocity at the center of each section. In slowly moving streams, determine the velocity by dropping a float attached to 5 feet of loose, monofilament fishing line. Measure the time required for the float to move 5 feet. For fast-moving streams, use a longer distance. Take at least three readings at each point. To calculate the flow rate from the information obtained above, use the following formula:

$$F = \frac{Ws \times D \times L \times C}{T}$$

where F = flow rate (cu. ft./sec.), Ws = surface width (ft.), D = mean depth (ft.), L = mean distance traveled by float (ft.), C = constant (0.8 for rough bottoms and 0.9 for smooth bottoms), and T = mean time for float (sec.).

For example, after using the above formula, you might have computed the stream's flow rate to be "10 cu. ft. per sec.".

3. **Calculation of Application Rate**

In order to calculate the application rate (expressed as "pound per sec"), you convert the rate in the table (expressed as "pound per acre-feet"), to "pound per cu. feet" and multiply by the flow rate (expressed as "cu. ft. per sec."). Depending on the size of the stream and the type of equipment, the rate could be expressed in other units, such as "ounces per hr."

The application rate for the stream above is calculated as follows:

$R_s = R_p \times C \times F$

where R_s = Application Rate for Stream (lb/sec), R_p = Application Rate for Pond (lb/acre feet), C = 1 acre foot/43560 cu. ft., and F = Flow Rate (cu. ft/sec).

In the example, the Application Rate for Stream would be:

R_s = 1 lb/0.74 acre-foot x 1 acre-foot/43560 cu. ft. x 10 cu. ft./sec.

R_s = .00031 lb/sec or 17.9 oz./hr.

4. **Exposure Time**

The "Exposure Time" would be the period of time (expressed in hours or seconds) during which target fish should **not** enter the lake or pond under treatment. In the example, this period of time could be 4 hours.

5. **Amount of Product**

Calculate the "Amount of Product" for a stream by multiplying the "Application Rate for Stream" by the "Exposure Time". In the example, the "Amount of Product" would be 71.6 oz. (17.9 oz./hr. x 4 hr.) or 4.5 lb.

RE-ENTRY STATEMENT

Do not allow swimming in rotenone-treated water until the application has been completed and all pesticide has been thoroughly mixed into the water according to labeling instructions.

12/98

APPENDIX I

BIOASSAY OF NOXFISH® (IN μg/L) TO FISH IN STANDARDIZED LABORATORY TESTS (MARKING AND BILLS 1976)

APPENDIX I

Bioassay of Noxfish® (in µg/L) to fish in standardized laboratory tests (Marking and Bills 1976).

Species	LC$_{50}$ and 95% confidence interval[1]	
	24 h	96 h
Bowfin	57.5	30.0
Amia calva	50.4–65.5	23.7–38.0
Coho salmon	71.6	62.0
Oncorhynchus kisutch	63–81.3	51.8–70.2
Chinook salmon	49.0	36.9
O. tshawytscha	44.3–54.2	33.9–40.2
Rainbow trout	68.9	46.0
O. mykiss	56.2–84.4	32.6–64.9
Atlantic salmon	35.0	21.5
Salmo salar	29.7–41.2	15.5–29.8
Brook trout	47.0	44.3
Salvelinus fontinalis	42.2–52.3	41.1–47.7
Lake trout	26.9	26.9
Salvelinus namaycush	19.8–36.5	19.8–36.5
Northern pike	44.9	33.0
Esox lucius	31.4–64.3	26.6–41.0
Goldfish		497
Carassius auratus		412–600
Common carp	84.0	50.0
Cyprinus carpio	74.7–94.4	41.1–60.8
Fathead minnow	400	142
Pimephales promelas	291–549	115–176
Longnose sucker	67.2	57.0
Catostomus catostomus	59.3–76.1	51.9–62.6
White sucker	71.9	68.0
Catostomus commersoni	64.0–80.8	51.0–85.6
Black bullhead	665	389
Ameiurus melas	516–856	298–507
Channel catfish	400	164
Ictalurus punctatus	234–684	138–196
Green sunfish	218	141
Lepomis cyanellus	197–241	114–174
Bluegill	149	141
Lepomis macrochirus	124–178	133–149
Smallmouth bass	93.2	79.0
Micropterus dolomieu	85.1–102	70.7–88.2
Largemouth bass	200	142
Micropterus salmoides	131–305	115–176
Yellow perch	92.0	70.0
Perca flavescens	80.1–106	59.8–82.0

1 Multiply values by 5% (0.05) to determine values as rotenone (in µg/L).

Appendix J

Bioassay Techniques

APPENDIX J

Bioassay Techniques

Tests are to be run using water and fish taken from the lake or stream being treated. The required amount of water is placed in plastic bags and rotenone added. Target concentrations will be predetermined from the lowest to the highest suggested range on the pesticide label. Three target fish, collected from the lake or stream, will be placed in plastic bags containing each target concentration. A control using untreated lake water will also be monitored.

The plastic bags should be 33 gal and at least 1.35-mil thickness. These bags must be gas permeable to pass oxygen from surrounding water. The bags are suspended from a rope or wire at the water surface. The amount of chemical needed for each bag is measured in a laboratory and brought to the site. Concentrations are measured in ppm of 5% (or 2.5% synergized) rotenone stock solution (concentrations are for total product not active ingredient).

Horton (1997) used Loeb and Engstrom-Heg's (1971) curve (Figure J.1) graphing concentration and time to loss of equilibrium of the fish to measure rotenone concentration. They used trout in the 6- to 9-in range and placed three in each cage to record times to loss of equilibrium. Loss of equilibrium is considered to have occurred when the mid-dorsal line comes in contact with the bottom of the container and the fish does not right itself and begin swimming normally again.

If loss of equilibrium occurs in less than 15 min at 18.5°C or 30 min at 8°C, Noxfish® concentration is fairly high (above 0.3 mg/L of formulation). You will get more accurate results by repeating the assay using a diluted sample, prepared by mixing 2 L of treated water with 8 L of untreated water. Concentrations of rotenone formulation can be estimated by finding the time to loss of equilibrium on the X-axis, connecting this with the formulation test temperature with a straight edge then reading the estimated concentration of rotenone formulation on the Y-axis (Figure J.1). The only way to know the exact concentration of rotenone is through analyzing water samples, but this method will provide an estimate.

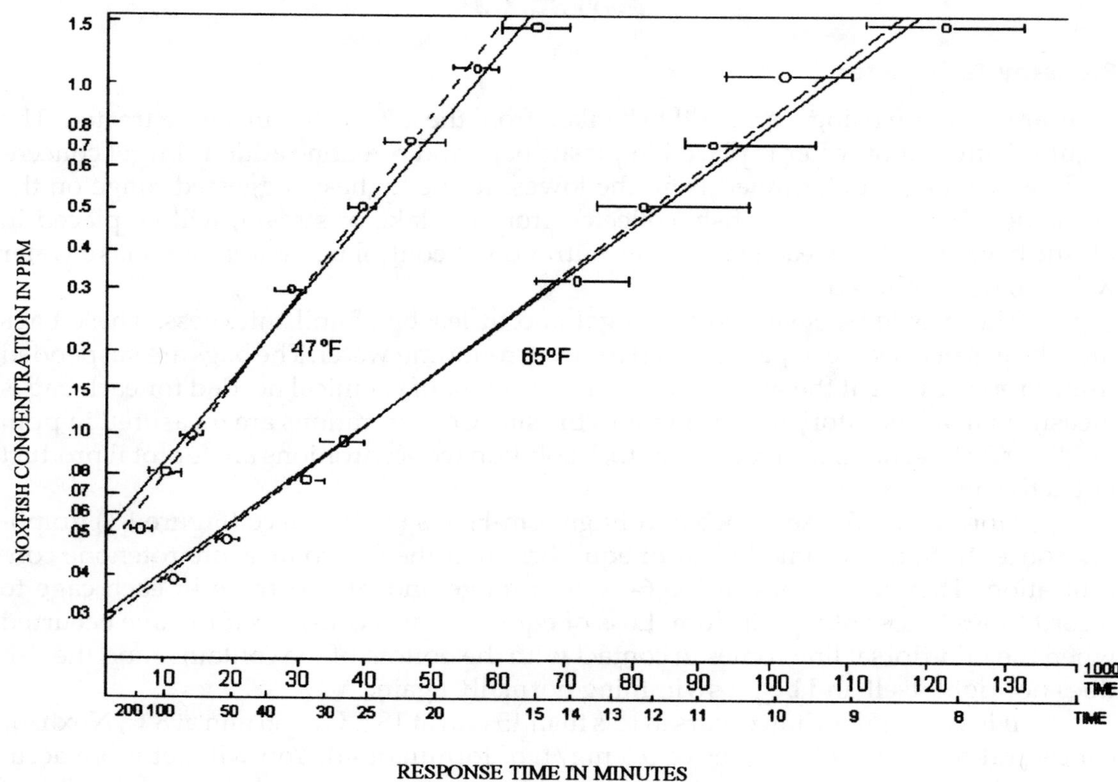

Figure J.1 Concentration-response curves for loss of equilibrium by brown trout *Salmo trutta* in Noxfish® dispersions. Solid lines represent semilogarithmic, dashed lines logarithmic, regressions. Each point represents mean time to loss of equilibrium + or − standard deviation. From Loeb and Engstrom-Heg (1971).

Appendix K

Liquid Application Equipment used to Treat Standing Water, as Described by California Department of Fish and Game (CDFG 1994)

APPENDIX K

Liquid application equipment used to treat standing water, as described by California Department of Fish and Game (1994).

When treating larger water bodies, a trash or sump pump of 4 to 6 horsepower with 2- to 3-in fittings is required. Intake hoses must be rigid, non-collapsible flex-pipe with metal basket end fittings. These are usually 8- to 12-ft in length and can be tied or weighted behind the boat.

Chemical barrels mounted in the boat are usually carried bung up and opened as needed. Draft tubes for the chemical are made of rigid, thin-walled electric conduit or any small (½- to ¾-in inside diameter) rigid, noncorrodible pipe. The tube is fitted with common rubber or vinyl hose with pipe clamps. This assembly is attached to a diffuser valve that is fitted to the base of the intake fitting on the pump. The diffuser valve is a ¾-in brass or galvanized faucet to control the flow of chemical to be mixed with the draft water at a 1:10 dilution. Testing with clean water is done to attain a 10% dilution rate as required on the pesticide label. A schematic diagram of the automatic dilution apparatus (allowing for closed-system application) is in Figure K.1.

The discharge hose is usually made of collapsible, vinyl material. Rotenone can be distributed from this hose in a fixed position below the gunnel of the boat or subsurface (when deep water treatment is necessary). The discharge hose can also be attached to a PVC manifold to conform to the needs of the project or to a high-velocity nozzle for spraying into inaccessible areas. The PVC manifold can be any configuration, but typically a short boom on either side of the boat will spread the chemical evenly. When smaller quantities (<200 gal) are used, small bilge or submersible 12-volt pumps are adequate. Diffusers and discharge plumbing are set up like the larger pump arrays with appropriate sizing.

Outboard motor-driven (propwash) venturi units may be effective for distribution on some projects. For bodies of water that are shallow but cover vast areas, aerial commercial applicators are often necessary. Typically, helicopters are more desirable because they can provide better accuracy and minimize drift.

Forklifts or cranes are required for lifting, loading, and unloading full barrels of rotenone (Figure K.2). Sets of barrels are placed on pallets for storage for mass transport. Individual barrels are lifted with a commercially available forklift-barrel adaptor. A boom truck or crane with a barrel clamp or barrel sling is used to lift and load individual barrels onto vehicles or boats. These are also commercially available.

Airboats and johnboats equipped with an air-cooled mud-king type motor are used to distribute liquid formulation into shallow or weeded areas inaccessible to convention boats and motors. The same water and rotenone containers are adapted to these boats.

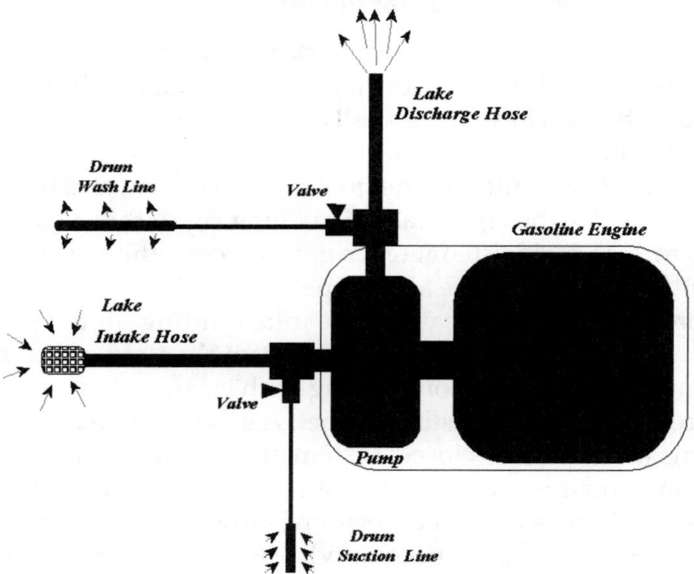

Figure K.1 Schematic diagram of automatic dilution apparatus for liquid rotenone using gasoline engine and pump. The apparatus allows for rinsing of rotenone barrels.

Figure K.2 Lifting rotenone drums with a crane onto a boat.

APPENDIX L

UTAH'S ROTENONE ASPIRATOR SYSTEM

APPENDIX L

Utah's rotenone aspirator system

The rotenone aspirator system developed by UDWR in 1990 used 3-in pipe and fittings. Smaller-sized pipe will work equally well, but will deliver less rotenone slurry per min. The UDWR was able to consistently mix and deliver 1,000 lbs of powder in 10 min with their system. The critical parts of the aspirator are a street elbow, a 1.25-in by 12-in nipple, and bell reducer (see Figure 3.2). A hole is cut in the back of the street elbow and the 1.25-in nipple slipped through this hole and slid forward until there is 0.125-in clearance between the nipple and the front of the bell reducer.

This forms the suction pipe for drawing rotenone powder from bags. The nipple is then welded into this position. The front of the nipple is ground until it forms a 45° angle from the inside edge out for a better fit. Threads in the front and inside of the bell housing may be ground smooth. The bell reducer and street elbow are marked so they can be taken apart and cleaned and rejoined in the same position. The clearance between the suction pipe and front of the bell housing is critical because water forced by the pressure pump through this small gap creates the vacuum on the powder suction line, and the violent action which mixes the powder and water.

When most of the powder is vacuumed from the container, the plastic container liner may be drawn into the suction hose. A valve placed on the suction hose or pipe near the entry into the street elbow can be opened to alleviate this problem. Opening the valve varying amounts will allow air to be drawn through the valve, reducing the pressure on the suction hose and allowing easier vacuuming of the last remaining powder from the bag. The small amount of powder remaining in the liner is rinsed into the lake if necessary.

Hose: Water and slurry delivery hose is made of 2- to 3-in diameter lightweight suction hose fitted with female camlok quick-release fitting on each end. Male camlok spools are used to connect lengths of hose together if necessary. This lightweight hose delivers fresh water from the reservoir to a high-pressure pump. High-pressure hose, rated at 100- to 150-lbs/in^2, is recommended to deliver water from the high-pressure pump to the aspirator, or the aspirator is coupled directly to the pump using pipe fittings. The powder suction hose is 2-in diameter rigid hose.

Pumps: A high-pressure pump, Gorman Rupp® series 60 centrifugal model with an enclosed impeller, or equivalent, is needed to pump water through the aspirator, creating the vacuum which draws the powder from the bag or container. This pump is close coupled to a Briggs and Stratton® Model 422437 twin cylinder, 18 HP, air-cooled gasoline engine, or equivalent. The pump is rated at 150 gal/min and 65 ft of head or greater.

Bags: Recent shipments of Pro-Noxfish® have come in 200-lb pressed paper (cardboard) barrels with plastic inner bags and Masonite lids secured with metal rings. The UDWR has completed projects using 50- and 1,000-lb bags (Strawberry Reservoir project) and 200-lb cardboard containers. The size of the project will dictate the size of container.

Special-sized containers must be ordered in advance and possibly supplied to the delivering company. The CDFG received powdered rotenone in 200-lb barrels for the Lake Davis project in 1997 (Figure L.1) and the UDWR used these barrels in a treatment of Navajo Lake in 1997 and Nine Mile Reservoir in 1998. These barrels proved to be an excellent package to handle. They were easy to handle without the use of heavy equipment and much more convenient than 50-lb bags. If larger bags are used, a boom truck or crane is needed to load them on boats.

Boats: In recent projects completed by UDWR, 16-ft to 18-ft johnboats were successfully used in place of larger barges. These smaller boats are easier to maneuver, can hold 600 to 800 lbs of rotenone, and can be operated by a two-person crew. They can be brought to shore at nearly any location on a lake or reservoir, and two workers can easily load 200-lb barrels on them by rolling the barrels up a plank or ramp and onto the boat.

Figure L.1 Application of powdered rotenone in 200-lb barrels in Lake Davis in 1997.

Appendix M

Nusyn-Noxfish® and Noxfish® Stream and River Use Monographs Calculation of Rotenone Concentration and Volume of use for Treatment of Flowing Water

NUSYN-NOXFISH® STREAM AND RIVER USE MONOGRAPH

USE IN STREAMS AND RIVERS

The following use directions are to provide guidance on how to make applications of Nusyn-Noxfish to streams and rivers. The unique nature of every application site could require minor adjustments to the method and rate of application. Should these unique conditions require major deviation from the use directions, a Special Local Need 24(c) registration should be obtained from the state.

Before applications of Nusyn-Noxfish can be made to streams and rivers, authorization must be obtained from state or federal Fish and Wildlife agencies. Since local environmental conditions will vary, consult with the state fish and wildlife agency to ensure the method and rate of application are appropriate for that site.

Contact the local water department to determine if any water intakes are within one mile downstream of the section of stream, river or canal to be treated. If so, coordinate the application with the water department to make sure the intakes are closed during treatment and detoxification.

Application Rates and Concentration of Rotenone

Slow Moving Rivers: In slow moving rivers and streams with little or no water exchange, use instructions for ponds, lakes and reservoirs.

Flowing Streams and Rivers: Apply rotenone as a drip for 4 to 8 hours to the flowing portion of the stream. Multiple application sites are used along the length of the treated stream, spaced approximately 1/2 to 2 miles apart depending on the water flow travel time between sites. Multiple sites are used because rotenone is diluted and detoxified with distance. Application sites are spaced at no more than 2 hours or at no less than 1 hour travel time intervals. This assures that the treated stream remains lethal to fish for a minimum of 2 hours. A non-toxic dye such as Rhodamine-WTR or fluorescein can be used to determine travel times. Cages containing live fish placed immediately upstream of the downstream application sites can be used as sentinels to assure that lethal conditions exist between sites.

Apply rotenone at each application site at a concentration of 0.5 to 2.0 parts per million of Nusyn-Noxfish. The amount of Nusyn-Noxfish needed at each site is dependent on stream flow (see Computation of Flow Rate for Stream).

Application of Undiluted Material

Nusyn-Noxfish can drain directly into the center of the stream at a rate of 0.85 to 3.4 cc per minute for each cubic foot per second of stream flow. Flow of undiluted Nusyn-Noxfish into the stream should be checked at least hourly. This is equivalent to from 0.5 to 2.0 ppm Nusyn-Noxfish, or from 0.012 to 0.050 ppm rotenone. Back-water, stagnant and spring areas of streams should be sprayed by hand with a 10% v/v solution of Nusyn-Noxfish in water to assure a complete coverage.

Calculation of Application Rate:

$$X = F(1.699\ B)$$

X = cc per minute of Nusyn-Noxfish applied to the stream, F = the flow rate (cu. ft./sec.) see Computation of Flow Rate for Stream section of the label, B = parts per million desired concentration of Nusyn-Noxfish.

Total Amount of Product Needed for Treatment: Streams should be treated for between 4 to 8 hours in order to clear the treated section of stream of fish. To determine the total amount of Nusyn-Noxfish required use the following equation;

$$Y = X(0.0158\ C)$$

Y = gallons of Nusyn-Noxfish required for the stream treatment, X = cc per minute of Nusyn-Noxfish applied to the stream, C = time in hours of the stream treatment.

Application of Diluted Material

Alternatively, for stream flows up to 25 cubic feet per second, continuous drip of diluted Nusyn-Noxfish at 80 cc per minute can be used. Flow of diluted Nusyn-Noxfish into the stream should be checked at least hourly. Use a 5 gallon reservoir over a 4 hour period, a 7.5 gallon reservoir over a 6 hour period, or a 10 gallon reservoir over an 8 hour period. The volume of the reservoir can be determined from the equation:

$$R = H \times 1.25$$

where R = the volume of the reservoir in gallons, and H = the duration of the application in hours.

The volume of Nusyn-Noxfish diluted with water in the reservoir is determined from the equation:

$$X = Y(102\ F)H$$

Where X = the cc of Nusyn-Noxfish diluted to 5 gallons, Y = parts per million desired concentration of Nusyn-Noxfish, F = the flow rate (cubic feet/second), H = the duration of the application (hours).

For flows over 25 cubic feet per second, additional reservoirs can be used concurrently. Back-water, stagnant and spring areas of streams should be sprayed by hand with a 10% v/v solution of Nusyn-Noxfish in water to assure a complete coverage.

Detoxification

To limit effects downstream, detoxification with potassium permanganate can be used at the downstream limit of the treated area. Within 1/2 to 2 miles of the furthest downstream Nusyn-Noxfish

NUSYN-NOXFISH®

application site, the rotenone can be detoxified with a potassium permanganate solution at a resultant stream concentration of 2 to 4 parts per million, depending on rotenone concentration and permanganate demand of the water. A 2.5% (10 pounds potassium permanganate to 50 gallons of water) permanganate solution is dripped in at a continuous rate using the equation:

$$X = Y(70\ F)$$

where X = cc of 2.5% permanganate solution per minute, Y = ppm of desired permanganate concentration, and F = cubic feet per second of stream flow.

Flow of permanganate should be checked at least hourly. Live fish in cages placed immediately above the permanganate application site will show signs of stress signaling the need for beginning detoxification. Detoxification can be terminated when replenished fish survive and show no signs of stress for at least four hours.

Detoxification of rotenone by permanganate requires between 15 to 30 minutes contact time (travel time). Cages containing live fish can be placed at these downstream intervals to judge the effectiveness of detoxification. At water temperatures of less than 50°F detoxification may be retarded, requiring a longer contact time.

Nusyn-Noxfish is a registered trademark of AgrEvo Environmental Health, Inc.

EPA Reg. No. 432-550

AgrEvo Environmental Health
95 Chestnut Ridge Road
Montvale, NJ 07645

NOXFISH®
STREAM AND RIVER USE MONOGRAPH

SPECIMEN LABEL

USE IN STREAMS AND RIVERS

The following use directions are to provide guidance on how to make applications of Noxfish to streams and rivers. The unique nature of every application site could require minor adjustments to the method and rate of application. Should these unique conditions require major deviation from the use directions, a Special Local Need 24(c) registration should be obtained from the state.

Before applications of Noxfish can be made to streams and rivers, authorization must be obtained from state or federal Fish and Wildlife agencies. Since local environmental conditions will vary, consult with the state Fish and Wildlife agency to ensure the method and rate of application are appropriate for that site.

Contact the local Water Department to determine if any water intakes are within one mile downstream of the section of stream, river or canal to be treated. If so, coordinate the application with the water department to make sure the intakes are closed during treatment and detoxification.

Application Rates and Concentration of Rotenone

Slow Moving Rivers: In slow moving rivers and streams with little or no water exchange, use instructions for ponds, lakes and reservoirs.

Flowing Streams and Rivers: Apply rotenone as a drip for 4 to 8 hours to the flowing portion of the stream. Multiple application sites are used along the length of the treated stream, spaced approximately 1/2 to 2 miles apart depending on the water flow travel time between sites. Multiple sites are used because rotenone is diluted and detoxified with distance. Application sites are spaced at no more than 2 hours or at no less than 1 hour travel time intervals. This assures that the treated stream remains lethal to fish for a minimum of 2 hours. A non-toxic dye such as Rhodamine-WTR or fluorescein can be used to determine travel times. Cages containing live fish placed immediately upstream of the downstream application sites can be used as sentinels to assure that lethal conditions exist between sites.

Apply rotenone at each application site at a concentration of 0.5 to 2.0 part per million of Noxfish. The amount of Noxfish needed at each site is dependent on stream flow (see Computation of Flow Rate for Stream).

Application of Undiluted Material

Noxfish can drain directly into the center of the stream at a rate of 0.85 to 3.4 cc per minute for each cubic foot per second of stream flow. Flow of undiluted Noxfish into the stream should be checked at least hourly. This is equivalent to from 0.5 to 2.0 ppm Noxfish, or from 0.025 to 0.100 ppm rotenone. Back-water, stagnant and spring areas of streams should be sprayed by hand with a 10% v/v solution of Noxfish in water to assure a complete coverage.

Calculation of Application Rate:

$$X = F(1.699\ B)$$

X = cc per minute of Noxfish applied to the stream, F = the flow rate (cu. ft./sec.) see Computation of Flow Rate for Stream section of the label, B = parts per million desired concentration of Noxfish.

Total Amount of Product Needed for Treatment: Streams should be treated for 4 to 8 hours in order to clear the treated section of stream of fish. To determine the total amount of Noxfish required use the following equation;

$$Y = X(0.0158\ C)$$

Y = gallons of Noxfish required for the stream treatment, X = cc per minute of Noxfish applied to the stream, C = time in hours of the stream treatment.

Application of Diluted Material

Alternatively, for stream flows up to 25 cubic feet per second, continuous drip of diluted Noxfish at 80 cc per minute can be used. Flow of diluted Noxfish into the stream should be checked at least hourly. Use a 5 gallon reservoir over a 4 hour period, a 7.5 gallon reservoir over a 6 hour period, or a 10 gallon reservoir over an 8 hour period. The volume of the reservoir can be determined from the equation:

$$R = H \times 1.25$$

where R = the volume of the reservoir in gallons, and H = the duration of the application in hours.

The volume of Noxfish diluted with water in the reservoir is determined from the equation:

$$X = Y(102\ F)H$$

where X = the cc of Noxfish diluted in the reservoir, Y = parts per million desired concentration of Noxfish, F = the flow rate (cubic feet/second), H = the duration of the application (hours).

For flows over 25 cubic feet per second, additional reservoirs can be used concurrently. Back-water, stagnant and spring areas of streams should be sprayed by hand with a 10% v/v solution of Noxfish in water to assure a complete coverage.

Detoxification

To limit effects downstream, detoxification with potassium permanganate can be used at the downstream limit of the treated area. Within 1/2 to 2 miles of the furthest downstream Noxfish application site, the rotenone can be detoxified with a potassium permanganate solution at a resultant stream concentration of 2 to

NOXFISH®

4 parts per million, depending on rotenone concentration and permanganate demand of the water. A 2.5% (10 pounds potassium permanganate to 50 gallons of water) permanganate solution is dripped in at a continuous rate using the equation:

$$X = Y(70\ F)$$

where X = cc of 2.5% permanganate solution per minute, Y = ppm of desired permanganate concentration, and F = cubic feet per second of stream flow.

Flow of permanganate should be checked at least hourly. Live fish in cages placed immediately above the permanganate application site will show signs of stress signaling the need for beginning detoxification. Detoxification can be terminated when replenished fish survive and show no signs of stress for at least four hours.

Detoxification of rotenone by permanganate requires between 15 to 30 minutes contact time (travel time). Cages containing live fish can be placed at these downstream intervals to judge the effectiveness of detoxification. At water temperature of less than 50°F detoxification may be retarded, requiring a longer contact time.

Noxfish is a registered trademark of AgrEvo Environmental Health, Inc.

AgrEvo Environmental Health
95 Chestnut Ridge Road
Montvale, NJ 07645

Appendix N

Drip Station Equipment

APPENDIX N

Drip Station Equipment

A drip station is established using a container which holds the measured amount of rotenone to provide the desired concentration and then filled with water to provide dilution and the proper delivery time.

The California Department of Fish and Game (CDFG 1994) describes a drip container that they have used successfully as consisting of a continuous flow siphon utilizing 5-gal can, two lengths of ¼-in copper tubing, brass compression fittings and a valve, and six ft of plastic ¼-in tubing. The copper tubes are placed through holes in the lids appropriate for the cans and soldered in place with sheet metal reinforcement for strength. The tubes are place so that the longer of the two will be ½-in from the can bottom when the cap is tightened, and the second will be ½-in shorter than the first. A valve is placed on the top end of the longer tube and the plastic tubing attached with a compression fitting. Cork or other gasket material is used inside the siphon lid to ensure the proper operation (see Figure 3.4a and 3.4b). This allows the pressure to come to equilibrium as the head in the container drops.

Drip containers used by Utah Division of Wildlife Resources consist of containers holding from 5 to 30 gal. Empty rotenone barrels are commonly used. These barrels come fitted with two openings in the top. One is approximately 1-in diameter, the other ½-in diameter. The smaller opening is fitted with a ½-in nipple to which a ½-in "T" is attached (see Figure 3.5). The "T" is fitted with a ½-in metal plug and stand pipe. A small hole is drilled in the bottom of the plug. The size of the hole will determine the rate of discharge. Utah Division of Wildlife Resources biologists have found that a 1/16-in hole will deliver about 50 mL/min and a 1/8-in hole will deliver about 200 mL/min. It is important to measure and clearly mark the delivery rate on each drip head, because holes drilled with a hand drill, even with the same drill bit, will deliver at slightly different rates.

Rotenone and water are added to the barrel through the larger opening and this opening is sealed tightly so no air may enter the barrel. The barrel is then positioned over the stream and tipped on its side.

Air enters the barrel through the "T" allowing pressure to come to equilibrium in the barrel. Flow rate will remain reasonably constant until the barrel is empty. A stand pipe can be attached to the elbow, but it is not necessary for successful operation. Barrels (30-gal) charged in this manner to full capacity with a 1/8-in drip opening will generally flow for approximately 9.5 h. Less chemical and water can be added to shorten discharge time.

Five-gal Jerry cans (U.S. military type containers) have also been adapted with the same drip heads for smaller projects or where equipment must be backpacked into an area. A ½-in hole is drilled in the bottom of the Jerry can and a ½-in nipple welded into the hole. A ½-in elbow is then attached and drilled to form the dripper. Containers fitted in this manner with a 1/8-in drip opening will dispense about 3.17 gal/h and must be recharged throughout the treatment period.

Appendix O

Rotenone–Sand Mixture used by Ground Crews to Treat Seeps and Springs, developed by Utah Division of Wildlife Resources in 1990

APPENDIX O

Rotenone-sand mixture used by ground crews to treat seeps and springs, developed by Utah Division of Wildlife Resources in 1990

During the treatment of Strawberry Reservoir tributaries, the Utah Division of Wildlife Resources (UDWR) developed a rotenone-sand mixture that proved superior to liquid applications (R. Spateholts, UDWR, personnel communication, 1990). The recipe for the mixture is 1 lb of powder rotenone to 1 lb of sand, and 2 oz of unflavored gelatin mixed with enough water to create a dough-like consistency. The gelatin holds the mixture together. It is then stored in 5-gal buckets with lids to keep the mixture moist. When this formulation was applied to seeps and springs at concentrations of 0.15, 0.25, and 0.35 mg/L, the active ingredient in the seeps and springs remained toxic for up to 12 h. Liquid formulations failed to kill some fish at the seeps and springs source. Ground crews were instructed to use one cup of rotenone sand mix to approximately 0.5 ft^3/s of water. This had to be monitored closely and crews often ran out of mix before completing their assigned area, indicating they were using too much mixture per seep and spring.

Appendix P

Neutralization Equipment and Techniques

APPENDIX P

Neutralization Equipment and Techniques

The Utah Division of Wildlife Resources uses bulk dispensing units to neutralize rotenone with $KMnO_4$ (C. Thompson, personal communication, 1999). These units (purchased from Acrison, Inc. in Moonachie, New Jersey) have a 2-cubic foot hopper that will hold about 150 lb of technical grade $KMnO_4$. They have an auger driven by an electric motor (Figure P.1) powered by a 115 VAC generator. The speed of the electric motor can be adjusted to dispense $KMnO_4$ at 2 to 45 lb/h. These units have performed in six major projects over seven years, are extremely reliable at dispensing the chemical at a constant rate, and substantially simplify the process of mixing the chemical with water. Mixing occurs in the stream. If a unit is placed at a rapids, falls, or whirlpool, complete mixing occurs in a short distance.

The following equipment checklist is used to assemble the necessary equipment prior to each neutralization treatment.

Personal gear (provided)
- 1 - Protective suit
- 1 - Pair rubber gloves
- 1 - Full-face respirator

Personal gear (to be provided by the individual)
- 1 - Cooler or thermos for drinks
- 1 - Pocket knife
- 1 - Pencil
- 1 - Flashlight
- 1 - Pair hip boots
- 1 - Watch with second calibration

Equipment (for each station)
- 1 - Volumetric dispenser
- 2 - Electrical extension cord
- 1 or 2 - Minnow traps
- 1 - 50-qt cooler for test fish
- 1 - Set of flood lights
- Log sheets
- Test fish/tank/cages
- 1 - Large screwdriver
- 1 - Pliers
- 1 - 9/16-in wrench
- 1 - Spark plug wrench
- 1 - Tarp
- 1 - Generator
- 1 - Gas can
- 1 - Radio
- 1 - Material Safety Data Sheet

Shared Equipment
- 1 - Backup Generator
- 1- Weighing scale
- 1 - Tape measure
- 1 - Flowmeter
- 1 - Potassium demand chemistry kit

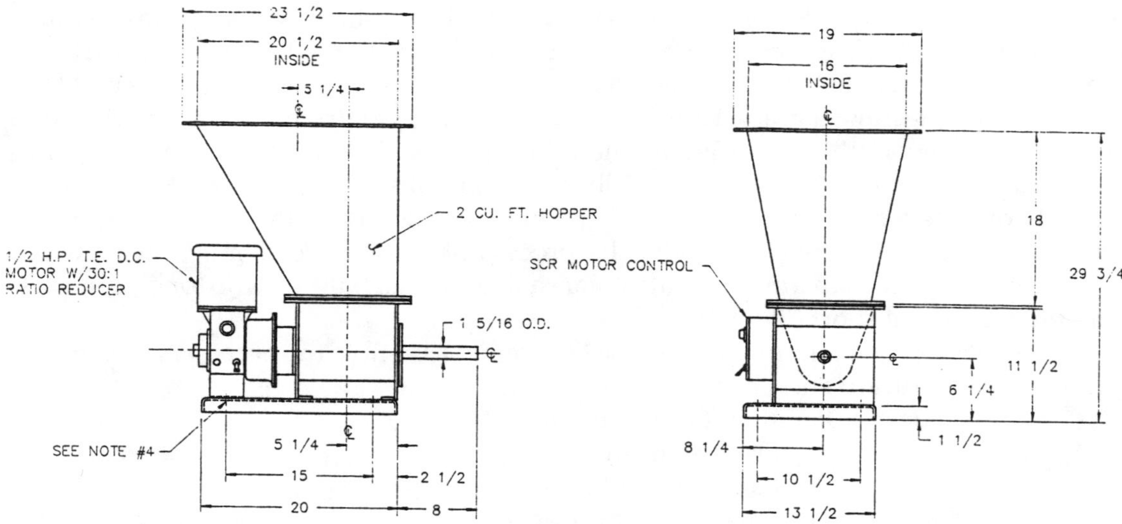

Figure P.1 Power auger dispenser used to dispense $KMnO_4$ to detoxify rotenone during chemical treatment projects to remove fish.

PROJECT ASSESSMENT 4

The American Fisheries Society (AFS) Fish Management Chemicals Subcommittee recommends project assessment on the basis of both short-term and long-term success and impacts. The extent of the assessment will be dictated by the objectives of the project and its complexity. For example, the treatment of a rearing facility may only require an assessment based on rearing success following the treatment. A fish sampling project may require only an assessment of the treatment success and neutralization (if any) procedures and a synopsis of the findings. Whole-lake and whole-stream treatments may require more extensive assessments. These requirements can be determined by good project planning.

The project plans (see Section 2.3) will specify much of the information needed in a project assessment. However, diligent record keeping during all phases of the project will be valuable to any project assessment. Assign personnel to keep records of various aspects of the project and make notes of pertinent details related to application of the chemical, responses of the biotic community, and success of any mitigation and neutralization efforts. This information will be important later in assessing the success of the project, responding to public inquiries, and defending against possible litigation. When conducting a project assessment, include an assessment of the overall effectiveness of the mitigation measures in lessening the impacts of the project on the environment. For most projects, a written report will be the end result of a project assessment.

4.1 SHORT-TERM ASSESSMENT

Analyze the immediate effectiveness of the treatment and any mitigation measures. Goals for the short-term assessment include (1) determination of the effectiveness of chemical application (i.e., distribution and neutralization of rotenone) and (2) recovery of baseline environmental conditions before stocking fish. Debrief all personnel as soon as the treatment phase of the project has been completed to identify problems, isolate causes, and propose corrective measures for future treatments. This effort involves the assessment of chemical and biological monitoring data and review of notes and observations recorded during and immediately following the treatment.

4.1.1 Assess the effectiveness of treatment and neutralization

Determine the effectiveness of the treatment and subsequent neutralization of rotenone. This assessment will enable project leaders to adjust plans based on the actual results of the treatment and neutralization operations. The effectiveness of these operations and related mitigation efforts can be judged by (1) counts of dead fish, (2) mortality of fish in live-cages, (3) bioassays, (4) sampling for the presence of live fish, (5) measurement of concentrations of rotenone in treatment and neutralized areas, and (6) visual observations. The sampling of baseline environmental conditions and estimates of dead fish from shoreline counts (static-water treatments) or collections from block nets (flowing-water treatments) are useful in evaluating the effectiveness of the treatment and neutralization. In flowing-water treatments, block nets and live-cages placed at various intervals downstream from detoxification stations are effective in determining the point at which total neutralization occurs and therefore the extent of the actual impact zone. Whenever using live-cages, it is always a good idea to hold a sample of fish in a nearby, comparable water body to rule out mortality that might be caused by factors other than rotenone. Standard fish sampling gear (e.g., electrofishing, gill nets, trap nets, fyke nets, seines) can be used to determine catch per unit effort after treatment and to evaluate success. Real time data can be used to adjust treatment and detoxification rates.

When the project objective is a partial treatment to reduce selected undesirable fish species, early efforts should be made to estimate the extent of the population reduction to determine if a repeat treatment is necessary. Where population sampling is the goal, it is desirable to be able to advise the fishing public when the fish community has returned to normal.

4.1.2 Recovery of baseline environmental conditions

If the project objective is the complete eradication of undesirable fish species, the recovery of baseline levels of target species is not warranted because the project goal was to reduce the target population below the baseline level. Therefore, recovery of baseline conditions for the target species should not occur if the project objectives are met.

However, if baseline levels of nontarget species and environmental conditions were evaluated before the treatment, evaluate these parameters after the treatment to determine if recovery objectives were met and if mitigation measures are needed. Survival and recovery of the aquatic community may be demonstrated by sampling plankton, macroinvertebrates (aquatic insects, crustacea, leeches, and mollusks), and amphibians (frogs, tadpoles, and larval and adult salamanders). Consideration should also be given to testing water quality, especially where public drinking water supplies are involved.

Before restocking, test the receiving water to determine if the piscicide has been sufficiently neutralized or has dissipated adequately to assure survival of stocked species. While chemical testing is possible, it is recommended that live-cages containing sensitive indicator species be used. Place the cages in representative locations and in areas where test fish will not be killed by stress or some event unrelated to the treatment (e.g., vandalism, predation, temperature). Be wary of embayments, bayous, or backwater or deepwater zones that may still harbor pockets of toxic water. In stratified lake environments, be sure cages are placed at depths where there is adequate oxygen for test fish to survive. In remote or wilderness areas where repeated field visits may not be feasible, it may be necessary to rely on the known rate of degradation of rotenone to predict when the water is detoxified. Formulas to predict natural degradation are contained in Post (1958) and Engstrom-Heg and Colesante (1979). When using formulas to calculate initial stocking dates, include a margin of safety.

4.1.3 Stock desired fish

As soon as testing indicates it is safe to restock, begin implementation of the restocking plan (see Section 2.1.5.8) or mitigation plan.

4.1.4 Written critique

This exercise may be the most significant part of the treatment, and if properly done will assist the agency in improving planning and implementation of future projects. Prepare a written summation and critique of the treatment as soon as possible after the treatment has been completed, in collaboration with your agency's legal advisors. Solicit input from all personnel involved in the treatment. It is advisable to have a meeting of all those involved in the treatment to get consensus on what worked and what could have been done differently. Each implemented plan from Section 2.3 should be assessed for accomplishing stated objectives. Determine if the plan was followed, what problems or issues were associated with the plan, and what improvements were needed. Send a draft of the written summation and critique to all personnel involved for review and consensus before completion of the final critique. When appropriate, use the critique to update policies and procedures.

4.2 LONG-TERM ASSESSMENT

Analyze the long-term impact of the project by evaluating the stated objectives over time. Goals for the long-term assessment include (1) determination of the duration of the treatment effectiveness and benefits, (2) determination of the success of the mitigation measures in lessening the impacts from the treatment, (3) assessment of the public perception of the success of the project, and (4) an overall assessment of the project.

4.2.1 Survival, growth, and harvest of fish

Monitor the fish community at regular intervals after the treatment when the project does not involve population sampling or annual treatment (e.g., rearing ponds). Periodic sampling can determine abundance, survival, age, and growth of desirable species and may verify the establishment of self-sustaining populations of valuable sport fish. Sampling will also provide continued information on other fish species, such as the reestablishment of the target fish species or the introduction of other fish, which may affect management goals.

When possible, collect angler use, catch, and harvest data to compare the actual benefits to predicted benefits. Attempts should be made to measure angler satisfaction through angler interviews, correspondence, and formal creel surveys. This information is extremely valuable when planning similar treatments on other waters.

4.2.2 Mitigation measures

Assess changes or alterations made to the original project that were adopted to mitigate or avoid significant adverse impacts on the environment. It is important that each change or alteration (to the original project) is judged for effectiveness, possible adverse impacts, and the need for mitigation. This assessment typically involves a monitoring program that specifies the implementation steps, timing, responsible party, and verification procedures for the various mitigation measures.

4.2.3 Public perception of treatment success

Assess how the angling and nonangling public perceived the project and address any public relations problems that may have occurred during the treatment. Consider publicizing the response of the fish community and angling success after the management effort so that anglers become aware of new and improved fishing opportunities. However, handle publication of this information carefully and in a manner that does not alienate property owners or create a situation where overexploitation may occur.

4.2.4 Overall assessment of project

Conduct an overall assessment of the project to document how well the original project objectives were met. Typically this should be a written report made within 2 to 5 years of the treatment. The fishery management plan should be updated after the outcome of the overall project assessment.

Issues and Responses 5

This section was written with the lay (nontechnical) public in mind with minimal use of technical terminology. It includes its own references for reproduction and distribution to the public independent of the remainder of the manual. The Fish Management Chemicals Subcommittee intends to update this information annually for access on the American Fisheries Society Web site.

5.1 General information

Q. What is rotenone?

A. Rotenone is a naturally occurring substance derived from the roots of tropical plants in the bean family Leguminosae including jewel vine *Derris* spp. and lacepod *Lonchocarpus* spp. Rotenone is very insoluble in water, and other materials can be added to disperse it throughout the water column in deep lakes and flowing waters. Rotenone is used either as a powder from ground-up plant roots (e.g., Pro-Noxfish®) or extracted from the roots and formulated as a liquid (e.g., Nusyn-Noxfish® and Noxfish®). The liquid formulations contain dispersants and emulsifiers (primarily naphthalene, methylnaphthalenes, and xylenes) that add little, if any, toxicity but disperse the rotenone throughout the water.

Q. How does rotenone work?

A. Rotenone does not suffocate fish or interfere with the uptake of oxygen in the blood as was long believed. Instead, it inhibits a biochemical process at the cellular level making it impossible for fish to use the oxygen absorbed in the blood and needed in the release of energy during respiration (Oberg 1967a, 1967b).

Q. Why is rotenone used in fish management?

A. Use of rotenone enables fisheries managers to eradicate entire populations and communities of fishes with minimum impact to nontarget wildlife. Following treatment, the desired population of fish is then reestablished in the water body. Although other approaches are useful as control measures, these are only partially effective in eradicating fish. Use of rotenone is the only sampling method that allows for an accurate estimation of standing crop (biomass of a population) of diverse fishes in large water bodies.

Q. Is rotenone a selective pesticide?

A. Although rotenone has some toxicity to all oxygen-breathing animals, it is selective to fish and other gill-breathing organisms at the concentrations used by fish biologists. In general, most common aquatic invertebrates are less sensitive than fish to rotenone. Some of the zooplankton (cladocerans and copepods) are equally sensitive; however, these do have life history stages that can survive the treatment. Snails and clams are quite tolerant. Shad, pike, trout, and salmon are among the most sensitive fish. Sunfish are less sensitive, and catfish are among the most tolerant (Marking and Bills 1976; Chandler and Marking 1982).

5.2 PUBLIC HEALTH

Q. Are there any public health effects from the use of rotenone?

A. Millions of dollars have been spent on research to determine the safety of rotenone before approval of use from the U.S. Environmental Protection Agency (USEPA). Much of this research has been directed toward potential effects on public health. This research has established that rotenone does not cause birth defects (Hazleton Raltech Laboratories 1982), reproductive dysfunction (Spencer and Sing 1982), gene mutations (Biotech Research 1981; Goethem et al. 1981; NAS 1983), or cancer (USEPA 1981b; Tisdel 1985). When used according to label instructions for the control of fish, rotenone poses little, if any, hazard to public health. The USEPA (1981b, 1989b) has concluded that the use of rotenone for fish control does not present a risk of unreasonable adverse effects to humans and the environment.

Q. What is a lifetime safe exposure level for rotenone?

A. The National Academy of Science (NAS 1983) has suggested a Suggested No-Adverse Response Level (SNARL) for rotenone in drinking water of 0.014 milligrams (mg) rotenone per liter of water (14 parts per billion [ppb]). The California Department of Health Services (memorandum from P. Berteau, California Department of Health Services, to B. Finlayson, California Department of Fish and Game, 26 June 1984) has suggested an Action Level (level of concern) for rotenone in drinking water of 0.004 mg rotenone per liter of water (4 ppb). These proposed life-time, allowable levels for drinking water are based on applying a 1,000-fold safety factor to the chronic feeding study of Ellis et al. (1980). These levels assume a lifetime of exposure to rotenone. For comparison, most rotenone treatments are done within the range of 0.025–0.25 mg rotenone per liter of water (25–250 ppb), and rotenone generally persists for no longer than a few weeks. In addition, rotenone treatments are only infrequently applied to any body of water.

Q. Is there any short-term danger associated with accidentally drinking rotenone-treated water?

A. The hazard associated with drinking water containing rotenone is very small because of the low concentration of rotenone used in the treatment (0.025–0.25 mg of rotenone per liter of water [25–250 ppb]) and the rapid breakdown of rotenone. Estimates on a single lethal dose to humans are 300–500 mg of rotenone per kilogram of body weight (Gleason et al. 1969). Hence, a 160-pound person would have to drink over 87,000 liters (23,000 gallons) of water treated at 0.25 mg of rotenone per liter of water (highest allowable treatment rate for fish management) at one sitting to receive a lethal dose; similarly, it is extremely unlikely that a 10-kilogram child would drink over 5,400 liters of water. An intake of 0.7 mg of rotenone per kilogram of body weight per day is considered safe (Haley 1978), far greater than the expected exposure resulting from the maximum fish management treatment rate of 0.25 mg of rotenone per liter of water.

Q. Can rotenone-treated water be used for public consumption or irrigation of crops?

A. Tolerances for rotenone in potable and irrigation water have not been established by USEPA, even though the studies required for setting tolerances have been completed. This does not mean that rotenone concentrations in drinking or irrigation waters will create problems, it just means that the USEPA has not established rotenone tolerances at the time of writing these guidelines. As a result, water containing residues of rotenone cannot be legally allowed for use as a domestic water source or on crops. During the treatment and for the period of time that rotenone residues are present, alternative water sources must be used for domestic and irrigation uses. Depending on initial rotenone concentration and environmental factors (e.g., temperature), this period can vary from 1 to 8 weeks (CDFG 1994; Finlayson and J. Harrington, unpublished data, presented at Chemical Rehabilitation Projects Symposium, Bozeman, Montana, 1991).

Q. Are there any risks to human health from materials in the liquid rotenone formulations?

A. The USEPA (1981b, 1989b) has concluded that the use of rotenone for fish control does not present a risk of unreasonable adverse effects to humans and the environment. The California Environmental Protection Agency found that adverse impacts from properly conducted, legal uses of liquid rotenone formulations in prescribed fish management projects were nonexistent or within acceptable levels (memorandum from J. Wells, California Department of Pesticide Regulation, to Finlayson, 3 August 1993). Liquid rotenone contains the carcinogen trichloroethylene (TCE). However, the TCE concentration in water immediately following treatment (less than 0.005 mg TCE per liter of water [5 ppb]) is within the level permissible in drinking water (0.005 mg TCE per liter of water; USEPA 1980b). None of the other materials including xylenes, naphthalene, piperonyl butoxide, and methylnaphthalenes exceed any water quality criteria or guidelines (based on lifetime exposure) set by the USEPA (1980a, 1981a, 1993). Many of these materials in the liquid rotenone formulations (trichloroethylene, naphthalene, and xylene) are the same as those found in fuel oil and are present in waters everywhere because of the frequent use of outboard motors.

Q. Is there any risk to public health from airborne rotenone?

A. No public health effects from rotenone use as a piscicide have been reported. The use of the powder Pro-Noxfish® and the liquid formulation Nusyn-Noxfish® have been monitored for airborne drift into adjacent areas. Airborne rotenone concentrations immediately adjacent to the treatment site, monitored in California during a treatment in 1997, varied from a high of 0.02 ppb rotenone (0.00053 mg of rotenone per cubic meter) immediately after application to nondetectable levels two weeks later (CARB 1997). The highest levels were approximately 1,000-fold lower than the estimated no observed effect level (NOEL) of 0.43 mg of rotenone per cubic meter of air for a 24-hour period estimated by the California Office of Environmental Health and Hazard Assessment and the California Department of Pesticide Regulation (CARB 1997). In the same monitoring program, TCE was detected only once at a trace amount in air at one spillway. The heavier hydrocarbons (naphthalene and methylnaphthalene) were found at 281 ppb (1.74 mg per cubic meter) in air immediately after treatment and diminished to 1.61 ppb (0.010 mg per cubic meter) in air within two weeks. Individuals can normally detect naphthalene and methylnaphthalene at levels between 40 and 84 ppb in the air. The highest levels of all materials in the 1997 monitoring program were found at a dam spillway because of water turbulence. The highest levels were determined not to be responsible for any health effects (CDPR 1998).

Q. How soon can people safely enter water treated with rotenone?

A. The USEPA (1981b) concluded that there was no reason to restrict the use of rotenone in waters intended for irrigation, livestock (with the possible exception of swine) consumption, and recreational swimming use. The USEPA (1990) ruled that a reentry interval was not needed for persons who swim in waters treated with rotenone based on an assessment of the toxicology data (e.g., skin, oral water intake) and exposure level. The reentry statement on the product labels—"do not swim in rotenone-treated water until the application has been completed and all the pesticide has been thoroughly mixed into the water according to labeling instructions"—indicates the safety of rotenone use for fish control. The reason for this slight waiting period is esthetic.

Q. Are people at risk from consuming fish stocked into a recently treated water body?

A. Fish are not stocked into a treated area until all of the toxic effects are gone and rotenone has dissipated. Hence, stocked fish cannot accumulate residues of rotenone from the water. Residues of rotenone in tolerant fish that survive a rotenone treatment won't last for more than several days because the bioaccumulation potential for rotenone is low and the half-life of rotenone in fish is about 1 day (Gingerich and Rach 1985; Gingerich 1986).

Q. Is there any risk to people from consuming fish that have been killed from rotenone?

A. The USEPA has not established guidelines for consuming fish killed with rotenone. Therefore, agencies cannot condone this practice. Additionally, there is a valid concern of risk of salmonella and other bacteriological poisoning from consuming fish that have been dead for a period of time. Fish that end up on land as a result of wave or wind action are no more a threat to public health than fish that die of natural causes.

5.3 ENVIRONMENTAL QUALITY

Q. Do dead and decaying fish pose any problems to the recovery of fishing?

A. Most dead fish will sink to the bottom of the treated body of water in several days, decompose, and release nutrients back into the water. These nutrients will directly stimulate phytoplankton and indirectly stimulate insect and zooplankton production. These organisms are a good food base for fish.

Q. Can the toxic effects of rotenone to fish and other aquatic life be neutralized?

A. In lakes or rivers, if biologists want to neutralize the effects of rotenone, potassium permanganate, an oxidizing agent, can be used. This is added to the water at a minimum 1:1 ratio with the concentration of rotenone applied plus sufficient additional permanganate to satisfy the oxygen demand caused by organic matter that may be present in the treated water. Neutralization of rotenone with permanganate may be impaired at water temperatures of 50°F (10°C) or less (CDFG 1994; AgrEvo, no date).

Q. What is the "pesticide" smell sometimes associated with the use of rotenone?

A. The aromatic smell (like the smell of mothballs) associated with the use of liquid rotenone formulations is likely airborne concentrations (greater than 40 ppb) of naphthalene and methylnaphthalene (CDPR 1998). This smell may last for several days, depending on air and water temperatures and wind direction. These relatively "heavy" organic compounds tend to sink (remain close to the ground) and move downwind. The California Department of Pesticide Regulation (CDPR 1998) found no health effects from this smell despite complaints.

Q. How long does rotenone persist?

A. The time for natural degradation (neutralization) of rotenone by hydrolysis is governed primarily by temperature. Studies in standing, ice-free waters in California show that rotenone completely degrades within

1 to 8 weeks within the temperature range of 10–20°C (CDFG 1994; Siepmann and Finlayson 1999; Finlayson and Harrington, unpublished); the estimated half-life values for California waters vary from 7.8 to 1.5 days, respectively. Other studies indicate half-life values of 13.9 hours to 10.3 days for water temperatures of 24°C and 5°C, respectively (Gilderhus et al. 1986, 1988). Rotenone dissipates in flowing waters relatively quickly (less than 24 hours) due to dilution and increased rates of hydrolysis (Borriston Laboratories 1983) and photolysis (Cheng et al. 1972; Biospherics 1982). Although rotenone can be found in lake sediments, the levels approximate those found in water, and breakdown of rotenone lags one to two weeks behind water levels. It is uncommon to find rotenone in stream sediments (CDFG 1994).

Q. How long do the materials other than rotenone persist from liquid formulation treatments?

A. Researchers in California have found other organic compounds associated with the use of the liquid formulation Nusyn-Noxfish® (CDFG 1994; Siepmann and Finlayson 1999; Finlayson and Harrington, unpublished). These include the volatile organic compounds (VOC) [xylene, trichlorethylene (TCE), toluene, and trimethylbenzene] and the semivolatile organic compounds (semiVOC) [piperonyl butoxide (PBO), naphthalene, 1-methyl naphthalene, and 2-methyl naphthalene] (Table 5.1). With the exception of PBO, the other organic compounds disappear before rotenone dissipates, typically within 1 to 3 weeks. Piperonyl butoxide, which is the other active ingredient (synergist) in Nusyn-Noxfish®, is relatively stable; photolysis does not contribute significantly to its degradation (Friedman and Epstein 1970). Piperonyl butoxide has persisted in deep lake waters at low temperatures (below 10°C) for approximately nine months. The VOC's do not accumulate in the sediment, and only naphthalene and the methyl naphthalenes temporarily (less than 8 weeks) accumulate in the sediments (CDFG 1994; Siepmann and Finlayson 1999; Finlayson and Harrington, unpublished).

Table 5.1. Persistence of rotenone and other organic compounds in water and sediment impoundments treated with 2 ppm rotenone formulation.

Compound	Initial water concentration (parts per billion)	Water persistence	Initial sediment concentration (parts per billion)	Sediment persistence
Rotenone	50	<8 weeks	522	<8 weeks
Trichloroethylene	1.4	<2 weeks	ND*	
Xylene	3.4	<2 weeks	ND	
Trimethylbenzene	0.68	<2 weeks	ND	
Naphthalene	140	<3 weeks	146	<8 weeks
1-m-naphthalene	150	<3 weeks	150	<4 weeks
2-m-naphthalene	340	<3 weeks	310	<4 weeks
Toluene	1.2	<2 weeks	ND	
Piperonyl Butoxide	30	<9 months	ND	

*ND=below detection limits

Q. Does the synergist piperonyl butoxide used in some formulations pose an environmental risk?

A. No, piperonyl butoxide has little toxicity to fish and wildlife and is not a risk to humans at the concentrations used in fish management (Roussel Bio Corporation, no date).

Q. Is rotenone likely to enter groundwater and pollute water supplies?

A. The ability of rotenone to move through soil is low to slight. Rotenone moves only 2 cm (<1 inch) in most types of soils. An exception would be in sandy soils where the movement is about 8 cm (slightly more than 3 inches). Rotenone is strongly bound to organic matter in soil so it is unlikely that rotenone would enter groundwater (Dawson et al. 1991). The other compounds in the liquid formulation Nusyn-Noxfish® have not been detected in groundwaters (CDFG 1994; Siepmann and Finlayson 1999; Finlayson and Harrington, unpublished).

Q. Are there any degradation products from rotenone that can cause environmental problems?

A. The metabolite of rotenone, rotenolone, persists longer than rotenone, especially in cold, alpine lakes (Finlayson and Harrington, unpublished). Rotenolone has been detected for as long as 6 weeks in cool water temperatures (<10°C) at high elevations (>8,000 feet). In part, this situation occurs because rotenone may be more susceptible to photolysis than rotenolone. However, studies have indicated that rotenolone is approximately one-tenth as lethal as rotenone (CDFG 1991a). In those rare cases of rotenolone persistence, fish stocking would be delayed until both rotenone and rotenolone residues have declined to nondetectable (<2 ppb) levels to err on the side of safety.

5.4 FISH AND WILDLIFE

Q. Does rotenone affect all aquatic animals the same?

A. No. Fish are more susceptible. All animals including fish, insects, birds, and mammals have natural enzymes in the digestive tract that neutralize rotenone, and the gastrointestinal absorption of rotenone is inefficient. However, fish (and some forms of amphibians and aquatic invertebrates) are more susceptible because rotenone is readily absorbed directly into their blood through their gills (non-oral route) and thus, digestive enzymes cannot neutralize it. Contrary to common belief, the other ingredients in Noxfish® and Nusyn-Noxfish® impart no toxicity to fish, insects, birds, or mammals (CDFG 1994). Rotenone residues in dead fish are generally very low (<0.1 ppm), unstable like those in water, and not readily absorbed through the gut of the animal eating fish.

Q. Will wildlife that eat dead fish and drink treated water be affected?

A. For the reasons listed above, birds and mammals that eat dead fish and drink treated water will not be affected. A bird weighing ¼ pound would have to consume 100 quarts of treated water or more than 40 pounds of fish and invertebrates within 24 hours to receive a lethal dose. This same bird would normally consume 0.2 ounces of water and 0.32 ounces of food daily; thus, a safety factor of 1,000- to 10,000-fold exists for birds and mammals. No latent or continuing toxicity is expected since under normal conditions rotenone will not persist for more than a few weeks (CDFG 1994).

Q. Will wildlife species be affected by the loss of their food supply following a rotenone treatment?

A. During recent treatments in California, fish-eating birds (i.e., herons and sea gulls) and mammals (i.e., raccoons) were found foraging on dying and recently dead fish for several days following treatment (CDFG 1994). Following this abundance of dead fish, a temporary reduction in food supplies for fish- or invertebrate-eating birds and mammals will result until the fish and invertebrates are restored. There is no indication that this temporary reduction results in any significant impacts to most bird or mammal populations because most animals can utilize other water bodies and sources for food. However, the temporary loss in food resources for sensitive animals during mating may cause unavoidable impacts. California has mitigated an impact to nesting bald eagles during mating by removing their eggs from the nest to an approved eagle recovery program out of the area (CDFG 1991b). Likewise, Michigan has mitigated the impacts to loons by delaying treatments until chicks have fledged.

Q. Is it safe for livestock to drink from rotenone-treated waters?

A. Rotenone was used for many years to control grubs on the backs of dairy and beef cattle. The USEPA (1981b) has stated that there is no need to restrict livestock consumption of treated waters. However, swine are more sensitive to rotenone than cattle (Thomson 1985).

5.5 REFERENCES

AgrEvo. No date. Nusyn-Noxfish stream and river monograph. AgrEvo Environmental Health, Montvale, New Jersey.

Biospherics. 1982. Aqueous photodegradation of ^{14}C-rotenone. Report to U.S. Geological Service, Upper Midwest Environmental Sciences Center (U.S. Fish and Wildlife Service Study 14-16-990-81-042), La Crosse, Wisconsin.

Biotech Research. 1981. Analytical studies for detection of chromosomal aberrations in fruit flies, rats, mice and horse bean. Report to U.S. Geological Service, Upper Midwest Environmental Sciences Center (U.S. Fish and Wildlife Service Study 14-16-990-80-54), La Crosse, Wisconsin.

Borriston Laboratories. 1983. Hydrolysis of ^{14}C-rotenone. Report to U.S. Geological Service, Upper Midwest Environmental Sciences Center (U.S. Fish and Wildlife Service Study), La Crosse, Wisconsin.

CARB (California Air Resources Board). 1997. Lake Davis fish kill emergency response—final report. CARB, Sacramento.

CDFG (California Department of Fish and Game). 1991a. Pesticide investigations unit, aquatic toxicology laboratory 1990 annual progress report. CDFG, Environmental Services Division, Sacramento.

CDFG (California Department of Fish and Game). 1991b. Northern pike eradication project - draft subsequent environmental impact report. CDFG, Inland Fisheries Division, Sacramento.

CDFG (California Department of Fish and Game). 1994. Rotenone use for fisheries management - final programmatic environmental impact report (SCH 92073015). CDFG, Environmental Services Division, Sacramento.

CDPR (California Department of Pesticide Regulation). 1998. A report on the illnesses related to the application of rotenone to Lake Davis. CDPR, Worker Health and Safety Branch, Report HS-1772, Sacramento.

Chandler, J., and L. Marking. 1982. Toxicity of rotenone to selected aquatic invertebrates and frog larvae. Progressive Fish-Culturist 44:78-80.

Cheng, H., I. Yamamoto, and J. Casida. 1972. Rotenone photodecomposition. Journal of Agricultural Food Chemistry 20:850-856.

Dawson, V. K., W. H. Gingerich, R. A. Davis, and P. A. Gilderhus. 1991. Rotenone persistence in freshwater ponds: effects of temperature and sediment adsorption. North American Journal of Fisheries Management 11:226-231.

Ellis, H., and six coauthors. 1980. Subchronic oral dosing study for safety evaluation of rotenone using dogs. Report to U.S. Geological Service, Upper Midwest Environmental Sciences Center (U.S. Fish and Wildlife Service Study 14-16-009-79-115), La Crosse, Wisconsin.

Friedman, M. A., and S. S. Epstein. 1970. Stability of piperonyl butoxide. Toxicology and Applied Pharmacology 17: 810-812.

Gilderhus, P. A., J. L. Allen, and V. K. Dawson. 1986. Persistence of rotenone in ponds at different temperatures. North American Journal of Fisheries Management 6:126-130.

Gilderhus, P. A., V. K. Dawson, and J. L. Allen. 1988. Deposition and persistence of rotenone in shallow ponds during cold and warm seasons. U.S. Fish and Wildlife Service, Investigations in Fish Control 95.

Gingerich, W. 1986. Tissue distribution and elimination of rotenone in rainbow trout. Aquatic Toxicology 8:27-40.

Gingerich, W., and J. Rach. 1985. Uptake, accumulation and depuration of ^{14}C-rotenone in bluegills (*Lepomis macrochirus*). Aquatic Toxicology 6:170-196.

Gleason, M., R. Gosselin, H. Hodge, and P. Smith. 1969. Clinical toxicology of commercial products. The William and Wilkins Company, Baltimore, Maryland.

Goethem, D., B. Barnhart, and S. Fotopoulos. 1981. Mutagenicity studies on rotenone. Report to U.S. Geological Service, Upper Midwest Environmental Sciences Center (U.S. Fish and Wildlife Service Study 14-16-009-80-076), La Crosse, Wisconsin.

Haley, T. 1978. A review of the literature of rotenone. Journal of Environmental Pathology and Toxicology 1:315-337.

Hazleton Raltech Laboratories. 1982. Teratology study with rotenone in rats. Report to U.S. Geological Service, Upper Midwest Environmental Sciences Center (U.S. Fish and Wildlife Service Study 81178), La Crosse, Wisconsin.

Marking, L., and T. Bills. 1976. Toxicity of rotenone to fish on standardized laboratory tests. U.S. Fish and Wildlife Service, Investigations in Fish Control, Bulletin 72.

NAS (National Academy of Science). 1983. Drinking water and health, volume 5. Safe Drinking Water Committee Board of Toxicology and Environmental Health Hazards, Commission on Life Sciences, National Research Council, National Academy Press, Washington, D.C.

Oberg, K. 1967a. On the principal way of attack of rotenone in fish. Archives for Zoology 18:217-220.

Oberg, K. 1967b. The reversibility of the respiration inhibition in gills and the ultrastructural changes in chloride cells from rotenone-poisoned marine teleost, *Gaduscallarius*. Experimental Cellular Research 45:590-602.

Roussel Bio Corporation. No date. Technical information sheet piperonyl butoxide insecticidal syngerist. Roussel Bio Corporation, Montvale, New Jersey.

Siepmann, S., and B. Finlayson. 1999. Chemical residues in water and sediment following rotenone application to Lake Davis, California. California Department of Fish and Game, Office of Spill Prevention and Response Administrative Report 99-2, Sacramento.

Spencer, F., and L. Sing. 1982. Reproductive responses to rotenone during decidualized pseudogestation and gestation in rats. Bulletin of Environmental Contamination and Toxicology 228:360-368.

Thomson, W. T. 1985. Agricultural chemicals, book 1: insecticides, acaracides and ovicides. Thomson Publications, Fresno, California.

Tisdel, M. 1985. Chronic toxicity study of rotenone in rats. Report to U.S. Geological Survey, Upper Midwest Environmental Sciences Center (U.S. Fish and Wildlife Service Study 6115-100), La Crosse, Wisconsin.

USEPA (U.S. Environmental Protection Agency). 1980a. Ambient water quality criteria for naphthalene. USEPA Document 440/5-80-059, Washington, D.C.

USEPA (U.S. Environmental Protection Agency). 1980b. Ambient water quality criteria trichloroethylene. USEPA Document 440/5-80-077, Washington, D.C.

USEPA (U.S. Environmental Protection Agency). 1981a. Advisory option for xylenes (dimethylbenzene). USEPA, Office of Drinking Water, Washington, D.C.

USEPA (U.S. Environmental Protection Agency). 1981b. Completion of pre-RPAR review of rotenone. USEPA, Office of Toxic Substances (June 22, 1981), Washington D.C.

USEPA (U.S. Environmental Protection Agency). 1989b. Guidance for the reregistration of pesticide products containing rotenone and associated resins as the active ingredient. USEPA Report 540/RS-89-039, Washington, D.C.

USEPA (U.S. Environmental Protection Agency). 1990. Rotenone re-entry statement for swimmers. USEPA Administrative 6704-Q, (January 17, 1990), Washington, D.C.

USEPA (U.S. Environmental Protection Agency). 1993. Water quality standards handbook. USEPA Report (EPA-823-B-93-002), Water Quality Standards Branch, Office of Science and Technology (September 1993), Washington, D.C.

References 6

AgrEvo. No date. Nusyn-Noxfish stream and river monograph. AgrEvo Environmental Health, Montvale, New Jersey.

Ball, R. C. 1948. A summary of experiments in Michigan lakes on the elimination of fish populations with rotenone, 1934–1942. Transactions of the American Fisheries Society 75:139–146.

Bettoli, P. W., and M. J. Maceina. 1996. Sampling with toxicants. Pages 303–333 *in* B. R. Murphy and D. W. Willis, editors. Fisheries techniques, second edition. American Fisheries Society, Bethesda, Maryland.

Biospherics. 1982. Aqueous photodegradation of ^{14}C-rotenone. Report to U.S. Geological Survey, Upper Midwest Environmental Sciences Center (U.S. Fish and Wildlife Service Study 14-16-990-81-042), La Crosse, Wisconsin.

Biotech Research. 1981. Analytical studies for detection of chromosomal aberrations in fruit flies, rats, mice and horse bean. Report to U.S. Geological Survey, Upper Midwest Environmental Sciences Center (U.S. Fish and Wildlife Service Study 14-16-990-80-54), La Crosse, Wisconsin.

Borriston Laboratories. 1983. Hydrolysis of ^{14}C-rotenone. Report to U.S. Geological Survey, Upper Midwest Environmental Sciences Center (U.S. Fish and Wildlife Service Study), La Crosse, Wisconsin.

CARB (California Air Resources Board). 1997. Lake Davis fish kill emergency response—final report. CARB, Sacramento.

CDFG (California Department of Fish and Game). 1991a. Pesticide investigations unit, aquatic toxicology laboratory 1990 annual progress report. CDFG, Environmental Services Division, Sacramento.

CDFG (California Department of Fish and Game). 1991b. Northern pike eradication project - draft subsequent environmental impact report. CDFG, Inland Fisheries Division, Sacramento.

CDFG (California Department of Fish and Game). 1994. Rotenone use for fisheries management - final programmatic environmental impact report (SCH 92073015). CDFG, Environmental Services Division, Sacramento.

CDPR (California Department of Pesticide Regulation). 1998. A report on the illnesses related to the application of rotenone to Lake Davis. CDPR, Worker Health and Safety Branch, Report HS-1772, Sacramento.

Chandler, J., and L. Marking. 1982. Toxicity of rotenone to selected aquatic invertebrates and frog larvae. Progressive Fish-Culturist 44:78-80.

Cheng, H., I. Yamamoto, and J. Casida. 1972. Rotenone photodecomposition. Journal of Agricultural Food Chemistry 20:850-856.

Cumming, K. B. 1975. History of fish toxicants in the United States. Pages 5-21 *in* P. H. Eschmeyer, editor. Rehabilitation of fish populations with toxicants: a symposium. American Fisheries Society, North Central Division, Special Publication 4, Bethesda, Maryland.

Davies, W. D., and W. L. Shelton. 1983. Sampling with toxicants. Pages 199-213 *in* L. A. Nielson and D. L. Johnson, editors. Fisheries techniques. American Fisheries Society, Bethesda, Maryland.

Dawson, V., P. Harmon, D. Schultz, and J. Allen. 1983. Rapid method of measuring rotenone in water at piscicidal concentrations. Transactions of the American Fisheries Society 112:725-728.

Dawson, V. K., W. H. Gingerich, R. A. Davis, and P. A. Gilderhus. 1991. Rotenone persistence in freshwater ponds: effects of temperature and sediment adsorption. North American Journal of Fisheries Management 11:226-231.

Ellis, H., and six coauthors. 1980. Subchronic oral dosing study for safety evaluation of rotenone using dogs. Report to U.S. Geological Survey, Upper Midwest Environmental Sciences Center (U.S. Fish and Wildlife Service Study 14-16-009-79-115), La Crosse, Wisconsin.

Engstrom-Heg, R. 1971. Direct measurement of potassium permanganate demand and residual potassium permanganate. New York Fish and Game Journal 18(2):117-122.

Engstrom-Heg, R. 1972. Kinetics of rotenone-potassium permanganate reactions as applied to the protection of trout streams. New York Fish and Game Journal 19(1):47-58.

Engstrom-Heg, R. 1976. Potassium permanganate demand of a stream bottom. New York Fish and Game Journal 23(2):155-159.

Engstrom-Heg, R., and R. T. Colesante. 1979. Predicting rotenone degradation in lakes and ponds. New York Fish and Game Journal 26(1):22-36.

Friedman, M. A., and S. S. Epstein. 1970. Stability of piperonyl butoxide. Toxicology and Applied Pharmacology 17:810-812.

Gallagher, A. S. 1999. Lake Morphology. Pages 165–174 *in* M. B. Bain and N. J. Stevenson, editors. Aquatic habitat assessment: common methods. American Fisheries Society, Bethesda, Maryland.

Gallagher, A. S., and N. J. Stevenson. 1999. Streamflow. Pages 149-158 *in* M. B. Bain and N. J. Stevenson, editors, Aquatic habitat assessment: common methods. American Fisheries Society, Bethesda, Maryland.

Gilderhus, P. A., J. L. Allen, and V. K. Dawson. 1986. Persistence of rotenone in ponds at different temperatures. North American Journal of Fisheries Management 6:129-130.

Gilderhus, P. A., V. K. Dawson, and J. L. Allen. 1988. Deposition and persistence of rotenone in shallow ponds during cold and warm seasons. U.S. Fish and Wildlife Service, Investigations in Fish Control 95.

Gingerich, W. 1986. Tissue distribution and elimination of rotenone in rainbow trout. Aquatic Toxicology 8:27-40.

Gingerich, W., and J. Rach. 1985. Uptake, accumulation and depuration of ^{14}C-rotenone in bluegills (*Lepomis macrochirus*). Aquatic Toxicology 6:170-196.

Gleason, M., R. Gosselin, H. Hodge, and P. Smith. 1969. Clinical toxicology of commercial products. The William and Wilkins Company, Baltimore, Maryland.

Goethem, D., B. Barnhart, and S. Fotopoulos. 1981. Mutagenicity studies on rotenone. Report to U.S. Geological Survey, Upper Midwest Environmental Sciences Center (U.S. Fish and Wildlife Service Study 14-16-009-80-076), La Crosse, Wisconsin.

Haley, T. 1978. A review of the literature of rotenone. Journal of Environmental Pathology and Toxicology 1:315-337.

Hazleton Raltech Laboratories. 1982. Teratology study with rotenone in rats. Report to U.S. Geological Survey, Upper Midwest Environmental Sciences Center (U.S. Fish and Wildlife Service Study 81178), La Crosse, Wisconsin.

Horton, W. D. 1997. Federal Aid in Sport Restoration fishery management program, lake renovation manual. Idaho Department of Fish and Game (IDFG 97-8) Boise.

Joint Subcommittee on Aquaculture, Working Group on Quality Assurance in Aquaculture Production. 1994. Guide to drug, vaccine, and pesticide use in aquaculture. Texas Agricultural Extension Service, B-5085, College Station.

Lennon, R. E., J. B. Hunn, R. A. Schnick, and R. M. Burress. 1970. Reclamation of ponds, lakes, and streams with fish toxicants: a review. Food and Agriculture Organization of the United Nations, Fisheries Technical Paper 100.

Loeb, H. A., and R. Engstrom-Heg. 1971. Estimation of rotenone concentration by bioassay. New York Fish and Game Journal 18(2):129-134.

Marking, L., and T. Bills. 1975. Toxicity of potassium permanganate to fish and its effectiveness for detoxifying antimycin. Transactions of the American Fisheries Society 104:579-583.

Marking, L., and T. Bills. 1976. Toxicity of rotenone to fish in standardized laboratory tests. U.S. Fish and Wildlife Service, Investigations in Fish Control, Bulletin 72.

MDNR (Michigan Department of Natural Resources). 1990. An assessment of human health and environmental effects of use of rotenone in Michigan's fisheries management programs. MDNR, Fisheries Division, Lansing.

MDNR (Michigan Department of Natural Resources). 1993. Policy and procedures of the use of piscicides and other compounds by the Fisheries Division in ponds, lakes, and streams. MDNR, Fisheries Division, Lansing.

M'Gonigle, R. H., and M. W. Smith. 1938. Cobequid hatchery - fish production in Second River and a new method of disease control. The Progressive Fish-Culturist 38:5–11.

NAS (National Academy of Science). 1983. Drinking water and health, volume 5. Safe Drinking Water Committee Board of Toxicology and Environmental Health Hazards, Commission on Life Sciences, National Research Council, National Academy Press, Washington, D.C.

NIFC (National Interagency Fire Center). 1994. Incident command system—national training curriculum modules. NIFC, Boise, Idaho.

Oberg, K. 1967a. On the principal way of attack of rotenone in fish. Archives for Zoology 18:217-220.

Oberg, K. 1967b. The reversibility of the respiration inhibition in gills and the ultrastructural changes in chloride cells from rotenone-poisoned marine teleost, *Gaduscallarius*. Experimental Cellular Research 45:590-602.

OSHA (Occupational Health and Safety Administration). 1978. Occupational health guidelines for rotenone. U.S. Department of Labor, Washington, D.C.

OSHA (Occupational Health and Safety Administration). 1988. Occupational safety and health guidelines for trichloroethylene, potential human carcinogen. U.S. Department of Health and Human Services, Washington, D.C.

Post, G. 1958. Time versus water temperature in rotenone dissipation. Proceedings of the Annual Conference Western Association of State Game and Fish Commissioners 48:279-284.

Province of British Columbia. 1993. Procedure manual: use of piscicides in fisheries management. Province of British Columbia, Ministry of Environment, Lands, and Parks.

Roussel Bio Corporation. No date. Technical information sheet piperonyl butoxide insecticidal syngerist. Roussel Bio Corporation, Montvale, New Jersey.

Sava, R. 1986. Guide to sampling air, water, soil and vegetation for chemical analysis. California Department of Food and Agriculture, Environmental Monitoring and Pest Management Branch, Sacramento.

Schnick, R. A. 1974. A review of the literature on the use of rotenone in fisheries. U.S. Fish and Wildlife Service, National Fishery Research Laboratory (NTIS PB-235 454), La Crosse, Wisconsin.

Siepmann, S., and B. Finlayson. 1999. Chemical residues in water and sediment following rotenone application to Lake Davis, California. California Department of Fish and Game, Office of Spill Prevention and Response Administrative Report 99-2, Sacramento.

Solman, V. E. F. 1950. History and use of fish poisons in the United States. Canadian Fish Culturist 8:3-16.

Sousa, R. J., F. P. Meyer, and R. A. Schnick. 1987a. Better fishing through management—how rotenone is used to help manage our fishery resources more effectively. U.S. Fish and Wildlife Service, Washington, D.C.

Sousa, R. J., F. P. Meyer, and R. A. Schnick. 1987b. Re-registration of rotenone: a state/federal cooperative effort. Fisheries 12(4):9-13.

Spencer, F., and L. Sing. 1982. Reproductive responses to rotenone during decidualized pseudogestation and gestation in rats. Bulletin of Environmental Contamination and Toxicology 228:360-368.

Thomson, W. T. 1985. Agricultural chemicals, book 1: insecticides, acaracides and ovicides. Thomson Publications, Fresno, California.

Tisdel, M. 1985. Chronic toxicity study of rotenone in rats. Report to U.S. Geological Survey, Upper Midwest Environmental Sciences Center (U.S. Fish and Wildlife Service Study No. 6115-100), La Crosse, Wisconsin.

USEPA (U.S. Environmental Protection Agency). 1973. Implementation plan, Pesticide Control Act. Federal Register 38(5):1142-1145.

USEPA (U.S. Environmental Protection Agency). 1980a. Ambient water quality criteria for naphthalene. USEPA Document 440/5-80-059, Washington, D.C.

USEPA (U.S. Environmental Protection Agency). 1980b. Ambient water quality criteria trichloroethylene. USEPA Report 440/5-80-077, Washington, D.C.

USEPA (U.S. Environmental Protection Agency). 1981a. Advisory option for xylenes (dimethylbenzene). USEPA, Office of Drinking Water, Washington, D.C.

USEPA (U.S. Environmental Protection Agency). 1981b. Completion of pre-RPAR review of rotenone. USEPA, Office of Toxic Substances (June 22, 1981), Washington D.C.

USEPA (U.S. Environmental Protection Agency). 1984a. Method 624: purgable organic compounds by gas chromatography/mass spectrometry (GC/MS). USEPA, Washington, D.C.

USEPA (U.S. Environmental Protection Agency). 1984b. Method 625: base/neutral and acid extractables by gas chromatography/mass spectrometry (GC/MS). USEPA, Washington, D.C.

USEPA (U.S. Environmental Protection Agency). 1986. Method 8310: polynuclear aromatic hydrocarbons. USEPA (EPA WS-846), Washington, D.C.

USEPA (U.S. Environmental Protection Agency). 1989a. Method 502.2: volatile organic compounds in water by purge and trap capillary column gas chromatography with photo-ionization and electrolytic conductivity detector in series. USEPA, Washington, D.C.

USEPA (U.S. Environmental Protection Agency). 1989b. Guidance for the reregistration of pesticide products containing rotenone and associated resins as the active ingredient. USEPA Report 540/RS-89-039, Washington, D.C.

USEPA (U.S. Environmental Protection Agency). 1990. Rotenone re-entry statement for swimmers. USEPA Administrative 6704-Q (January 17, 1990), Washington, D.C.

USEPA (U.S. Environmental Protection Agency). 1993. Water quality standards handbook. USEPA Report (EPA-823-B-93-002), Water Quality Standards Branch, Office of Science and Technology (September 1993), Washington, D.C.

USEPA (U.S. Environmental Protection Agency). 1994a. Method 8270B: semi-volatile organic compounds by gas chromatography/mass spectrometry (GC/MS): capillary column technique. USEPA, Washington, D.C.

USEPA (U.S. Environmental Protection Agency). 1994b. Method 8260B: volatile organic compounds by gas chromatography/mass spectrometry (GC/MS): capillary column technique. USEPA, Washington, D.C.

WDG (Washington Department of Game). 1986. Rotenone and trout stocking. WDG, Fisheries Management Division, Report 86-2, Olympia.

WDW (Washington Department of Wildlife). 1992. Environmental impact statement lake and stream rehabilitations, 1992-1993—final supplemental report. WDW, Habitat and Fisheries Management Divisions, Report 92-14, Olympia.